SpringerBriefs in Physics

For further volumes:
http://www.springer.com/series/8902

Percival McCormack

Vortex, Molecular Spin and Nanovorticity

An Introduction

 Springer

Percival McCormack
Department of BioEngineering
University of Illinois at Chicago
218 SEO, 851 S. Morgan St. (M/C 063)
Chicago, IL 60607–7052
USA
pmccorm@uic.edu

ISSN 2191-5423   e-ISSN 2191-5431
ISBN 978-1-4614-0256-5   e-ISBN 978-1-4614-0257-2
DOI 10.1007/978-1-4614-0257-2
Springer New York Dordrecht Heidelberg London

Library of Congress Control Number: 2011936742

Printed on acid-free paper

Springer is part of Springer Science+Business Media (www.springer.com)

Contents

Chapter 1
The Vortex

1.1 Brief History

The first notable works on fluid motions appeared in the early seventeenth century. Descartes in his "Principia Philosophiae" (1644) considered that space was filled with frictional vortices, so that the planets are carried along by the vortex motions. Vortices had fascinated mankind for many centuries before Descartes. It was considered that life had started in the water of the primeval vortex – in whirlwinds and whirlpools.

The oldest description of a tidal vortex was given by Homer (eighth century BC). The returning heroes of the Odyssey had to face the danger of the giant whirlpool Charybde [1]. The Celts and Teutons also believed that life was created in this whirlpool. The connection between cosmic whirl and the Maelstrom is described by de Santillana [2].

Empedocles (492–432 BC) lived in Sicily and believed in four primary substances – earth, water, air, and fire. He also distinguished attractive and repelling "'forces' as 'love' and 'strife'", respectively. He demonstrated the centrifugal effect by the retention of fluid within a rotating, liquid-filled ladle. He used this effect to explain the position of the celestial bodies in the firmament.

Anaxagoras (500–428 BC) did not believe in the basic four elements but thought of matter as a continuum. He extended the vortex concept to the total world process [3]. With hindsight, today it would appear that this vortex concept of Anaxagoras may be close to a universal truth. It is interesting to note that Plato and his disciple Aristotle (384–322 BC) rejected the vortex theory of the micro- and macro-cosmos. However, in his book "Meteorologica" and in the pseudo-Aristotelian scripts "Problemata Mechanica" and "De Mundo", vortices *are* described for their own sake. The cause, occurrence, and motion of whirlwinds and tidal vortices are considered.

In Roman times, meteorological vortices were described by Seneca (0–65 AD), Pliny (23–79 AD), and Lucan (39–65 AD). In the Middle Ages, the early development of a scientific terminology appeared [4, 5].

P. McCormack, *Vortex, Molecular Spin and Nanovorticity: An Introduction*,
SpringerBriefs in Physics, DOI 10.1007/978-1-4614-0257-2_1,
© Percival McCormack 2012

In the thirteenth century, Gerard of Brussels studied the kinematics of rotating bodies – he conceptualized solid-body rotation (see his text – "Liber de motu"). Nicole Oresme (circa 1325–1382) studied the commensurability of rotating celestial bodies. He identified the return of a single body along a circular orbit from an arbitrary point to the same point, as a circulation, whereas the return of several bodies from an arbitrary initial state to exactly the same state (commensurable), or to a similar state (incommensurable), is a revolution.

The Renaissance period saw the climax in the use of art to describe vortex motions, and leading this effort was Leonardo da Vinci (1452–1515). He perceived the vortex and the wave as the manifestation of power and motion, and used spiral and wavy elements in his paintings. He pioneered the concept of turbulent motion; identified the difference between the potential vortex and solid-body rotation; and studied channel vortex flow and vortex formation in the wake of obstacles [6].

1.1.1 Cartesian Vortex Theory (Sixteenth and Seventeenth Centuries)

In this period, Kepler found that the planetary orbits are elliptic not circular. He attributed this distortion to magnetic attraction and repulsion. He conceived of a magnetic vortex caused by the rotation of the sun – the essence of Cartesian vortex theory [7, 8]. The theory is based on the assumption that matter has extension and is identified with space (it is a distributed system). The implication was that there is no vacuum and bodies interact by direct contact. When a body moves the surrounding fluid, particles are induced into motion around it. This is valid for a continuum, but not for celestial bodies. The problem was resolved in 1687 by Newton in his "Principia" – he advocated that material bodies interact over a distance by gravitational attraction. The success of Newtonian mechanics in describing planetary motion overwhelmed all opposition. The eighteenth and nineteenth centuries saw the birth and development of classical mechanics – based upon Newton's theory.

Hadley in 1735 published a theory of the general circulation of the atmosphere. Although since rejected, his theory was the start of a long effort to develop a satisfactory model of atmospheric motions [9]. The effort still goes on today with the aid of high-speed computers. In 1749, Boscovitch published a comprehensive analysis of tornadoes [10]. Kant in 1755 proposed that the sun and planets developed out of a rotating gaseous cloud.

The foundation of fluid mechanics is attributed to Euler [11]. He derived the equations of motion for an inviscid fluid. He used a mathematical term that later was identified as the "vorticity vector." This led to the development of the equations of motion for viscous fluids – known as the Navier–Stokes equations [12]. The analytical treatment of vortex motion started with Herman von Helmholtz's classic paper "On Integrals of the Hydrodynamic Equations Corresponding to Vortex Motion" in 1858. Kelvin in 1869 demonstrated the necessity of Helmholtz's theorems for the existence of vortex motions and developed a circulation theorem named after him. These works formed the basis for the modern vortex theory. They also led to the extension of the vortex concept into other areas of physics.

Based on the argument that magnetic fields are rotatory, Maxwell in 1861 used the vortex model in his electromagnetic theory. The equivalence of electric and magnetic fields was inherent in this theory and was pointed out by Boltzmann in 1891. During the latter half of the nineteenth century, Lord Kelvin developed a theory of the properties of atoms and molecules based on vortex rings [13–15]. However, vortices in nature are unstable and decay and these properties were incompatible with the atomic model. Although the attempt to interpret the structure of the physical world in terms of a mechanical vortex model failed, the vortex concept and its applications in classical mechanics, fluid, and aerodynamics, and now in the fields of superfluid physics and superconductivity, have been very successful.

1.1.2 The Twentieth Century

The vortex theorems of Helmholtz and Lord Kelvin were further developed by Crocco (1937), Ertel (1942), and Vazsonyi (1945) – see [16]. The Taylor–Proudman theorem has made an important contribution to the analysis of rotating fluid systems [17]. Benard, von Karman, Taylor, and Gortler have identified various vortex configurations, all connected with the development of flow instability. For example, Taylor–Gortler vortices develop in curved flow (over a concave plate) and involve a conversion from two-dimensional disturbances into a three-dimensional configuration [18]. The onset of flow instability is the prelude to transition to turbulence and early research on turbulence is identified with the work of Boussinesq, Taylor, Prandtl, Heisenberg, and Reynolds [19, 20].

The *dynamics* of turbulent flow is based on the hypothesis that the Navier–Stokes equations are valid for turbulent flow. But, the *mechanics*, or structure, of turbulence is a *statistical* mechanics. Turbulent boundary layers are complicated by the fact that their structure is neither homogeneous nor isotropic. Little progress has been made to date on the *theory* of nonisotropic turbulence. Considerable progress has been made, however, in the study of at least locally isotropic turbulence. This has been facilitated by the use of computational fluid dynamics (CFD).

In 1941, a general theory of locally isotropic turbulence was formulated by Kolmogorov which predicted a number of laws governing turbulent flow for large Reynolds numbers. The fundamental physical concepts which form the basis of Kolmogorov's theory can be summarized as follows:

1. A turbulent flow at large Reynolds numbers is considered to be the result of the superposing of disturbances (vortices or eddies) of all possible sizes. Only the very largest of these vortices are due directly to the instability of the mean flow.

2. The motion of the large vortices is unstable and this produces smaller (secondary) vortices: the latter produce third-order vortices and so on.
3. The motion of the smallest vortices is "laminar" and depends basically on the molecular viscosity.
4. The motion of all but the largest vortices can be assumed to be homogeneous and isotropic.
5. The motion of all vortices whose scales are much smaller than the *local* structure of the flow must be subject to general statistical laws which do not depend on the geometry of the flow or on the properties of the mean flow.

The establishment of these general statistical laws constitutes the theory of local isotropic turbulence.

In aero- and hydro-dynamics, Lanchester made a unique contribution by suggesting that a vortex acting as an aerofoil is the cause of lift.

The bound vortex theory of lift has been developed by many scientists, but of special note are the names of Prandtl et al. [21]. Analytical design of airplanes, propellers, and turbines rapidly followed.

In meteorology, theoretical weather prediction is based on Bjerknes circulation theorem and Rossby's solution of the vorticity transport equation. The fact that to understand global atmospheric circulation it is essential to know the occurrence and migration of cyclonic and anticyclonic air masses in mid-latitudes, which was discovered by Jeffreys in 1926 [9]. The great contribution of Rotunno in the 1980s to tornado dynamics must also be noted [22]. In 1908, Benard related the vortex concept to sound and this led to the theory of aerodynamic sound generation based on vortical motion in an unsteady flow. Lighthill [23] and Powell [24] have been pioneers in this area, and Sarpkaya [25] in the closely related area of vortex-induced vibration of bodies.

In astronomy and astrophysical dynamics, the vortex phenomenon is common. Hubble identified spiral nebulae as galaxies in the mid-1920s, and in 1943 Weizsacher developed a vortex theory for the generation of the planetary system. In recent decades, studies of exotic stars have been made. Neutron stars with enormously high angular velocities have been postulated and discovered, as well as rotating black holes whose existence and characteristics require the general theory of relativity and quantum mechanics [26]. Magnificent images of the atmospheric vortex arrays from Jupiter have been transmitted back to Earth by Voyagers 1 and 2 [27, 28]. Many unsolved vortex problems remain, such as the motion of planetary atmospheres and convection flows inside stars [29, 30].

With the development of ultra-fast computers with great storage capacity, the ease of programming, and associated graphics capability, it has become possible to solve numerically (even in real time) the basic equations of fluid motion for nontrivial problems. The discipline of computational fluid dynamics (CFD) is now recognized in its own right [31, 32]. For example, numerical solution of the vorticity transport equation was the key to understanding unsteady flows [33]. Finite element, finite difference methods, and extensive meshes with moving frames of reference have facilitated the design of aerofoils, engine intakes, and nacelles in turbulent hypersonic flow.

When the underlying discretization is of the vorticity field, rather than the velocity field, the numerical method for approximating the solution of the incompressible Euler form of the Navier–Stokes equations is known as the vortex method [34, 35]. Leonard [36, 37] has studied a 3D version of the vortex-in-cell method and computed 3D flows.

Advances in the fields of turbulence and unionized and ionized compressible and incompressible flow – all necessitating a statistical mechanical approach – will probably be "rate controlled" by advances in CFD techniques.

In the 1940s, it was found that the peculiar, or superfluid, properties of helium 2 (fountain effect, second sound, and frictionless flow) could be largely accounted for on the basis of the phenomenological two-fluid theory of Tisza [38] and Landau [39]. In 1954, London [40] proposed that below the so-called "λ-point" (a discontinuity in the specific heat versus temperature curve), helium is a quantum fluid whose essential feature is the macroscopic occupation of a single quantum state. Excitations in liquid helium were identified as phonons (similar to those in solids) and rotons, which correspond to possible rotational modes of motion in the liquid. It was Onsager in 1950 [41] who showed that there was a possible wave function for the liquid, which would produce a motion analogous to classical vortex motion, and suggested the possible existence of quantized vortices in helium 2. Feynmann [42] in 1955 developed this conjecture into a theory which has been successful in explaining many of the peculiar phenomena observed in the superfluid. Experimental and theoretical studies of the superfluid state continue today, with the fascinating extension of the quantized vortex to the superconducting state and the interaction with, and generation of, magnetic fields [43, 44]. Moreover, as vortices are described by solutions of field equations, the topology of vortices is now relevant to elementary particle physics [45].

Vortex physics is now well developed, from the microscopic, or atomic, world to the astronomical world. The vortex concept spans an enormous size range from the truly microscopic to the truly macroscopic (10^{-8}–10^{18} cm – a scale factor of 10^{26}!). The significance of the fact that vortex physics spans this enormous range, unchanged apart from the quantum restriction at the microscopic level, remains to be elucidated.

1.2 Kinematics

1.2.1 Definition of a Vortex

A vortex is defined as the motion of fluid particles around a central spin axis. The individual particle paths may be circular, or not – see Fig. 1.1. If the paths are the same in every plane normal to the axis of rotation, one has a cylindrical

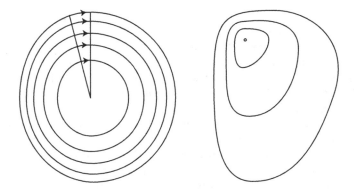

Fig. 1.1

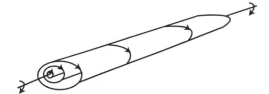

Fig. 1.2

Fig. 1.3

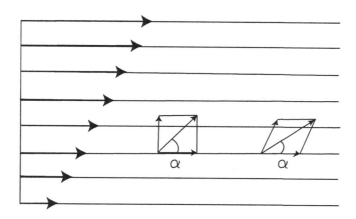

Fig. 1.4

vortex – see Fig. 1.2. Many natural vortices have pathlines that are not perpendicular to the axis of rotation, but are oblique to it. A particular case is the spiral vortex, illustrated in Fig. 1.3. Rather being based on pathlines, the *flow pattern* at a point in space may be used to define the vortex. This is the basis of the Cauchy and Stokes definition [46]. They called the angular velocity of a fluid at a point in space "vorticity" or spin. Thus, vortex motion involves a basic mode of rotation, along with translational and deformational motion.

Other terms for vortex, such as eddy, circulation, spiral, whirl, and cyclone, are used and differ only in their physical, geophysical, or engineering meaning.

All motion of a fluid within a finite space surrounded by a solid boundary must result in rotation. This follows from the conservation of matter. Such rotational motion is also possible in an infinite space, given that the fluid far from the source of the motion is at rest. As mentioned earlier, the angular velocity about a point is spin. Galaxies and virtually zero mass particles all spin. Spin is governed by the conservation of angular momentum – a fact that is fundamental to many analyses in physics. In the atmosphere, in space, and in the ocean, if a specific volume of fluid that is rotating is compacted, the rate of rotation must increase to conserve angular momentum.

It must be noted that the presence of vorticity, or spin, is basic to any vortex, but one can have vorticity and *no vortex*, as in the shear flow illustrated in Fig. 1.4.

1.2.2 Vorticity and Circulation

Stokes (1845) showed for viscous fluids and vonHelmholtz for ideal fluids that the motion of deformable bodies, such as fluids, can be broken down into a sum of a translation, a rotation, and a strain. The strain may be further divided into a pure linear strain and a pure shearing (angular) strain.

In fluid media, the parameter of prime interest is the rate of deformation, or velocity of the fluid. In vector notation, the velocity at a point x is given by

$$\mathbf{v} = \mathbf{v}_0 + (\mathbf{r} - \mathbf{r}_0)x\boldsymbol{\omega} + (\mathbf{r} - \mathbf{r}_0). \tag{1.1}$$

The first term of (1.1), u_0, v_0, w_0, is a translatory contribution or a displacement rate.

The second term, which can be written indicially as

$$\delta v_i^{(q)} = -\left(\frac{1}{2}\right)\varepsilon_{ijk}r_j\omega_k. \tag{1.2}$$

is the antisymmetric contribution to the change in velocity and

$$
\begin{aligned}
2\omega_1 &= \frac{\partial v_2}{\partial r_2} - \frac{\partial v_2}{\partial r_1} \\
2\omega_2 &= \frac{\partial v_1}{\partial r_3} - \frac{\partial v_3}{\partial r_1} \\
2\omega_3 &= \frac{\partial v_2}{\partial r_1} - \frac{\partial v_1}{\partial r_2}.
\end{aligned}
\tag{1.3}
$$

In rectangular Cartesian coordinates

$$\omega = \left(\frac{1}{2}\right)\left[\left(\frac{\partial w}{\partial y} - \frac{\partial v}{\partial z}\right)\mathbf{i} + \left(\frac{\partial u}{\partial z} - \frac{\partial w}{\partial x}\right)\mathbf{j} + \left(\frac{\partial v}{\partial x} - \frac{\partial u}{\partial y}\right)\mathbf{k}\right]. \tag{1.4}$$

$\delta v^{(q)}$ is the velocity produced at position r relative to a point about which there is rigid-body rotation with angular velocity ω where

$$\boldsymbol{\zeta} = 2\omega = \nabla \times \mathbf{v}. \tag{1.5}$$

The vector $\boldsymbol{\zeta}$ is called the local vorticity (or spin) of the fluid.

The third component of (1.1) involves the deformation, or strain, of the fluid, both linear (extensional) and shear. The strain is described in terms of the strain rate tensor $T = e_{ij}$ defined by the square array

$$
T =
\begin{matrix}
e_{xx} & e_{xy} & e_{xz} \\
e_{yx} & e_{yy} & e_{yz} \\
e_{zx} & e_{zy} & e_{zz}
\end{matrix}
$$

$$
=
\begin{matrix}
\frac{\partial u}{\partial x} & \frac{1}{2}\left(\frac{\partial u}{\partial y} + \frac{\partial v}{\partial x}\right) & \frac{1}{2}\left(\frac{\partial u}{\partial z} + \frac{\partial w}{\partial x}\right) \\
\frac{1}{2}\left(\frac{\partial v}{\partial x} + \frac{\partial u}{\partial y}\right) & \frac{\partial v}{\partial y} & \frac{1}{2}\left(\frac{\partial v}{\partial z} + \frac{\partial w}{\partial y}\right) \\
\frac{1}{2}\left(\frac{\partial w}{\partial x} + \frac{\partial u}{\partial z}\right) & \frac{1}{2}\left(\frac{\partial w}{\partial y} + \frac{\partial v}{\partial z}\right) & \frac{\partial w}{\partial z}
\end{matrix}
. \tag{1.6}
$$

The sum of the diagonal terms (extensional strain rates), or the trace of the tensor, is the divergence of the velocity, or the fluid dilation θ:

$$\theta = e_1 + e_2 + e_3 = \frac{\partial u}{\partial x} + \frac{\partial v}{\partial y} + \frac{\partial w}{\partial z} = \nabla.\mathbf{v}. \tag{1.7}$$

Kelvin in 1869 introduced the concept of circulation. The circulation around any closed curve (or circuit) in the fluid is *defined* by the integral

$$\Gamma = \int_c v.dl = \int v_i dl_i. \tag{1.8}$$

Now

$$\boldsymbol{\zeta} = \nabla \times \omega. \tag{1.9}$$

ζ is a function of position in the fluid and represents, at each point, twice the angular velocity of a fluid element. If ζ_n is the component of vorticity normal to the surface S bounded by the circuit C, then

$$\Gamma = \int_s \zeta_n dS, \tag{1.10}$$

and

$\zeta_n = \mathbf{n} \cdot (\nabla \times \mathbf{v})$ where $\mathbf{n}$ is the unit normal vector. The relation between the circulation and vorticity is thus given by

$$\Gamma = \int_s (\mathbf{n}.\zeta)dS = 2 \int_s (\mathbf{n}.\omega)dS, \tag{1.11}$$

where $\mathbf{n}.\omega = (1/2)d\Gamma/dS$

When the fluid flow is such that the fluid elements do not rotate ($\zeta = 0$), the circulation will be zero for any closed circuit in the flow region.

It must be noted that while vorticity is a vector quantity and deals with the spin of a fluid particle, circulation is a scalar quantity and deals with fluid rotation over a finite area of fluid. The tornado phenomenon can be used to contrast these two measures of rotation. Consider a tornado with no, or negligible, vertical velocity, a core of radius a, and rotating at a constant angular velocity ω. Within the core, the velocity is given by

$$v_c = \omega r (r < a).$$

Outside the core, where $r > a$, we have a free vortex and $v_o = k/r$ – see (1.12) below – where k is the vortex strength.

In (1.5) above, the components of vorticity are

$$\zeta_x = \frac{\partial w}{\partial y} - \frac{\partial v}{\partial z},$$

$$\zeta_y = \frac{\partial u}{\partial z} - \frac{\partial w}{\partial x},$$

$$\zeta_z = \frac{\partial v}{\partial x} - \frac{\partial u}{\partial y}.$$

Outside of the core of the tornado, there is no vorticity and all these components will be zero. By assuming continuity of velocity at $r = a$, the vortex strength, k, can be evaluated:

$$v_c = v_0 \text{ at } r = a \text{ and } \omega a = \frac{k}{a},$$

therefore

$$k = \omega a^2.$$

Inside the core

$$\boldsymbol{\zeta} = \nabla \times \mathbf{v} = \nabla \times (\boldsymbol{\omega} \times \mathbf{r}).$$

Now

$$\boldsymbol{\omega} = 0\mathbf{i} + 0\mathbf{j} + \omega\mathbf{k},$$

$$\mathbf{r} = x\mathbf{i} + y\mathbf{j} + 0\mathbf{k},$$

$$\therefore \boldsymbol{\omega} x \mathbf{r} = -y\mathbf{i} + x\mathbf{j},$$

and

$$\nabla \times \mathbf{v} = 2\omega = \boldsymbol{\zeta}.$$

The *circulation* within the core is given by

$$\Gamma = \int v.dl = \int^{2\pi} v_c r d\theta = \int^{2\pi} (\omega r) r d\theta = 2\pi\omega r^2.$$

Thus, within the core the circulation is dependent on the value of r, increasing with distance from the center of the vortex. Now, the circulation is equal to the product of the vorticity and the area, and, as the vorticity is constant in the core,

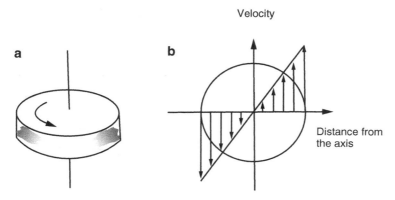

Fig. 1.5

the circulation is determined solely by the area (or the radius). The circulation can be regarded as the product of the vorticity and the area, and so as $\zeta = 2\omega$, the circulation outside the core is just

$$\Gamma = 2\pi\omega a^2.$$

Since both a and ω are constant, the circulation remains constant and independent of the distance from the center of the vortex. The entire contribution to circulation in the region of the free vortex originates in the core of the tornado.

1.2.3   Plane Circular Vortex

This vortex involves solid-body rotation, with the velocity of the fluid particles increasing linearly with distance from the center of rotation – see Fig. 1.5. A uniform translational motion can be superposed on the solid-body rotation and then the fluid particles will travel along helical paths – see Fig. 1.6.

By contrast, consider a circular rod rotating in a fluid at constant angular velocity – see Fig. 1.7. The fluid velocity is a maximum and equal to the angular velocity of the rod at the rod's surface (no-slip condition). With increasing distance from the rod, the fluid velocity decreases linearly. Such a fluid is called a "potential vortex" and has no vorticity *except* at the center of rotation. The potential vortex, in other words, has zero vorticity outside of the core.

When the fluid velocity depends only on the distance from the center of rotation, r, in cylindrical coordinates (u_r, u_θ, u_z) then observations (as first noted by Leonardo da Vinci) show that

$$Vr = \text{constant}, \tag{1.12}$$

Fig. 1.6

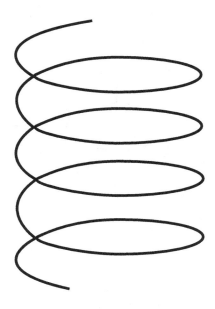

or

$$u_\theta = \frac{\text{constant}}{r}.$$

The flow field defined by (1.12) is called the free vortex. The constant is called the vortex strength. It must be noted that the velocity approaches infinity at $r = 0$ – a singularity exists in the flow field. This led to the necessity for the vortex core, where solid-body rotation occurs and the velocity is *zero* at the center of rotation (see Fig. 1.7). The free vortex is a good approximation to the velocity in a bathtub vortex, a tornado, or a hurricane.

For solid-body rotation then, the flow equations are

$$\frac{V}{r} = \text{constant} = \frac{d\theta}{dt}, \tag{1.13}$$

where $d\theta/dt$ is the angular rotation rate – see Fig. 1.8. This flow is called a *forced vortex* and occurs in the *steady* flow of the fluid inside a cylinder rotating at a constant rate. The frictional force from the rotating bottom slowly transmits through the fluid till it rotates as a solid body.

In a similar way, one component of the Earth's spin produces an effective rotation of the Earth's surface and this forces a vortex flow from the surface into the atmosphere. The rotation is one revolution every 24 h at the poles and zero at the equator.

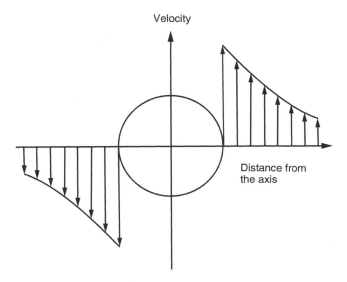

Fig. 1.7

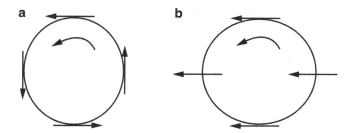

Fig. 1.8

This is the source of the rotation of cyclones, hurricanes, and tornados. The component of the Earth's spin changes at the equator, and so the vortex rotation is counterclockwise in the Northern Hemisphere and clockwise in the Southern Hemisphere.

It is obvious that the term "vortex" can be associated with quite different velocity distributions.

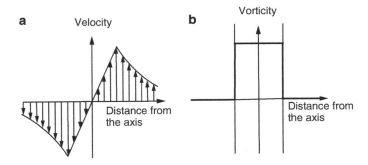

Fig. 1.9

1.2.4 Rankine Vortex

The combination of the rotational core and surrounding irrotational ("spin-less") region is known as the Rankine combined vortex – the overall velocity distribution is shown in Fig. 1.9. The effect of viscosity is to cause the vorticity to diffuse outward and the core to increase in size. The discontinuity in the slope of the velocity at $r = a$ is also smoothed out. While the diameter of the core of the vortex can be determined exactly, the definition of the diameter of the vortex is arbitrary – for example, it could be defined as the distance from the core center to the point at which the velocity has decreased to, say, 10% of the maximum velocity.

If the radius of the vortex core reduces to a point, one gets a "vortex line" in which the whole flow field is free of spin (or vorticity) except at the axis of rotation where it is infinite, but the circulation remains finite. A bundle of vortex lines forms a "vortex tube."

Since there are no shear forces in solid-body rotation, no energy is required to maintain the motion. In contrast, the rotating rod must continually provide energy to the potential vortex to allow for the loss of energy by shearing of the fluid particles. If solid-body rotation were to occur in a vacuum, once the rotation is started no further input of energy is required to maintain it.

1.2.4.1 Pressure Distribution in the Combined Vortex

For a rectilinear vortex, the tangential velocity in the irrotational region *outside* the core, and based on (1.8), is given by

$$q_\theta = \frac{\Gamma}{2\pi r},$$

(1.14)

where $q_\theta = (u^2 + w^2)^{1/2}$ and u, w are the x and y components of velocity, respectively, Γ is the vortex strength, and r is the radial distance from the center of the core.

Using this in Bernouilli's equation, one obtains

$$p + \left(\frac{\rho}{2}\right)q^2 + \rho g h = p + \rho g h + \left(\frac{\rho}{2}\right)\left(\frac{\Gamma^2}{4\pi^2 r^2}\right) = \Pi, \tag{1.15}$$

where Π is Bernouilli's constant.

Taking conditions far from the core as a reference ($r = \infty$) designated by the subscript 0

$$(p + \rho g h) - (p + \rho g h)_0 = -\frac{(\rho/8\pi^2)\Gamma^2}{r^2}. \tag{1.16}$$

The pressure varies as the inverse radius squared.

In the rotational core, as vorticity exists, then the Bernouilli's constant will vary with r. The more basic Euler equation must now be resorted to. From the Euler equations, the acceleration component normal to the streamlines is given by

$$a_n = \frac{\partial q}{\partial t} + \frac{q^2}{n} = \frac{-(1/\rho)\partial}{\partial n(p + \rho g h)}. \tag{1.17}$$

For current streamlines (as in a vortex core), $n = r$, and for steady flow, therefore,

$$\frac{\partial(p + \rho g h)}{\partial r} = -\frac{\rho q_\theta^2}{r}. \tag{1.18}$$

Now

$$q_\theta = \frac{(\Gamma/2\pi a)r}{a},$$

and so

$$\frac{\partial(p + \rho g h)}{\partial r} = \frac{\rho q_\theta^2}{r} = \frac{\rho r \Gamma^2}{4\pi^2 a^4}, \tag{1.19}$$

where a is the radius of the core.

Integration of this equation gives the pressure relation

$$p + \rho g h = \left(\frac{\rho}{2}\right)\left(\frac{\Gamma^2}{4\pi^2 a^2}\right)\left(\frac{r}{a}\right)^2 + C. \tag{1.20}$$

Fig. 1.10

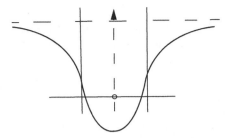

The pressure rises as the square of the radius. The constant C is evaluated at $r = a$, so that

$$C = (p + \rho gh)_0 - \frac{\rho \Gamma^2}{4\pi^2 a^2}.$$

Therefore,

$$(p + \rho gh) - (p + \rho gh)_0 = \left(\frac{\rho}{2}\right)\left(\frac{\Gamma^2}{4\pi^2 a^2}\right)\left[\left(\frac{r}{a}\right)^2 - 2\right]. \tag{1.21}$$

The pressure variation in the core of the combined, or Rankine, vortex, given by (1.21) and the pressure variation in the surrounding irrotational region, is sketched in Fig. 1.10.

The pressure at the center of a water vortex can decrease to such an extent that the dissolved gases in the water separate so that the water evaporates even at room temperature. This results in gas or air cavities around the rotation axis – cavitation occurs (a hollow core forms).

1.2.5   Curved Vortex Lines, Tubes, and Vortex Rings

In Cartesian coordinates, the three components of vorticity are

$$\omega_x = \frac{\partial w}{\partial y} - \frac{\partial v}{\partial z} \quad \omega_y = \frac{\partial u}{\partial z} - \frac{\partial w}{\partial x} \quad \omega_z = \frac{\partial v}{\partial x} - \frac{\partial u}{\partial y}, \tag{1.22}$$

and

$$\frac{\partial \omega_x}{\partial x} = \frac{\partial^2 w}{\partial y \partial x} - \frac{\partial^2 v}{\partial z \partial x}, \tag{1.23}$$

$$\frac{\partial \omega_y}{\partial y} = \frac{\partial^2 u}{\partial z \partial y} - \frac{\partial^2 w}{\partial x \partial y}, \tag{1.24}$$

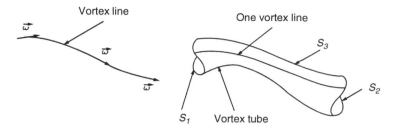

Fig. 1.11

$$\frac{\partial \omega_z}{\partial z} = \frac{\partial^2 v}{\partial x \partial z} - \frac{\partial^2 u}{\partial y \partial z}. \tag{1.25}$$

As the two components on the right-hand side of each of these equations are equal

$$\nabla \cdot \boldsymbol{\omega} = \frac{\partial \omega_x}{\partial x} + \frac{\partial \omega_y}{\partial y} + \frac{\partial \omega_z}{\partial z} = 0. \tag{1.26}$$

Thus, vorticity components are related in the same way as the velocity components in incompressible flow. Integrating (1.26) over a finite volume and applying the divergence theorem:

$$\int_V \nabla \cdot \boldsymbol{\omega} dV = 0,$$

and hence

$$\int_S \boldsymbol{\omega} \cdot \mathbf{n} dS = 0. \tag{1.27}$$

where S is the surface bounding V and $\mathbf{n}$ is the outward normal to S. The *vortex line* is defined as a line that is everywhere tangent to the local vorticity vector (cf. the streamline). A *vortex tube* is the set of vortex lines passing through a simply connected surface in space. Figure 1.11 sketches a vortex line and a vortex tube. Obviously, $\boldsymbol{\omega} \cdot \mathbf{n} = 0$ on the surface S_3 of the tube. Applying (1.27) to the tube, then

$$\int_{S1} \boldsymbol{\omega} \cdot \mathbf{n} dS + \int_{S2} \boldsymbol{\omega} \cdot \mathbf{n} dS = 0. \tag{1.28}$$

From the definition of circulation, the first integral in (1.28) is $-\Gamma_1$ and the second is Γ_2 (the outward normals are in opposite directions), so that (1.28) becomes

$$\Gamma_1 = \Gamma_2. \tag{1.29}$$

Fig. 1.12

Thus, the circulation around a vortex tube is constant (Helmholtz vortex law I). This law dictates that a vortex tube can never terminate in a fluid. Vortex tubes are thus constrained to either forming loops within a fluid or terminating at a solid boundary or fluid–fluid interface. Vortex lines which curve have a velocity component perpendicular to the plane of curvature.

1.2.5.1   Biot–Savart Law

The fluid velocity induced by an arbitrary distribution of vorticity is

$$
\begin{aligned}
\mathbf{q}_V &= \left(\frac{1}{4\pi}\right) \int_V \left[\nabla \times \frac{\zeta(x',y',z')}{r}\right] dV' \\
&\equiv \left(\frac{1}{4\pi}\right) \int_V \left[\boldsymbol{\zeta}(x',y',z') \times \frac{\mathbf{r}}{r^3}\right] dV',
\end{aligned}
\tag{1.30}
$$

and $\int_V ()\,dV'$ is a volume integral ($dV' = dx'\,dy'\,dz'$) throughout the volume in which the vorticity is distributed. The local induced velocity at a point P due to an element $\zeta\,ds$ is mutually perpendicular to $\mathbf{r}$ and $\boldsymbol{\zeta}$ in the direction determined by the right-hand rule by rotating $\boldsymbol{\zeta}$ into $\mathbf{r}$. The velocity at P is obtained by integrating around the vortex filament. Thus, for the vorticity distributed over a closed curve, see Fig. 1.12.

$$
\mathbf{q}_V = \left(\frac{1}{4\pi}\right) \int_s \left[\frac{\zeta(x',y',z') \times \mathbf{r}}{r^3}\right] ds.
\tag{1.31}
$$

This is called the *Biot–Savart law* and was deduced experimentally by Biot and Savart in 1820 as the magnetic vector induced by a steady electric current flowing in a closed conductor.

Under the conditions that,

$$
\nabla \times \mathbf{q}_V = \boldsymbol{\omega},
$$
$$
\nabla \times \mathbf{q}_V = 0 \text{(dilation is zero everywhere)},
$$
$$
\mathbf{w} \times \mathbf{n} \text{ is zero at each point of the boundary,}
$$

$$
\mathbf{q}_V = \left(\frac{-1}{4\pi}\right) \int \left[\frac{\mathbf{r} \times \omega}{r^3}\right] dV,
\tag{1.32}
$$

(see [47], McCormack and Crane, p. 138).

Fig. 1.13

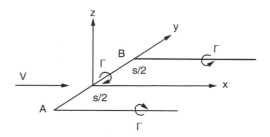

This equation will determine the velocity field associated with the line singularity formed by a vortex tube contracting on to a curve with the vortex strength remaining constant and equal to (say) Γ. Now, if δs is an elemental vector length of the line vortex lying in the volume δV, then

$$\int_{\delta V} \zeta \, dV = \Gamma \delta s,$$

where ζ is the local vorticity vector and

$$\mathbf{q}_V = \left(\frac{-\Gamma}{4\pi}\right) \int \mathbf{r} \times \frac{ds(\mathbf{r})}{r^3}. \tag{1.33}$$

The corresponding Biot–Savart law in electromagnetic theory for a steady current passing around a closed conducting loop is

$$\mathbf{H} = I \int \mathbf{r} \times ds(\mathbf{r})/r^3.$$

The field strength $\mathbf{H}$ is analogous to the fluid velocity and the current to the vortex strength.

Consider the horseshoe vortex of circulation Γ, representing a wing of finite span, with the bound vortex line A–B of length s, and two semi-infinite tip vortices emanating from A and B (Fig. 1.13). The velocity fields at a point x, y, z (see [48] – von Karman and Burgers) are

Bound vortex: $(q_x)_B$; $(q_y)_B = 0$; $(q_z)_B$
Tip vortices: $(q_x)_T = 0$; $(q_y)_T$; $(q_z)_T$ and

$$
\begin{aligned}
q_x &= (q_x)_B, \\
q_y &= (q_y)_T, \\
q_z &= (q_z)_B + (q_z)_T,
\end{aligned} \tag{1.34}
$$

where

$$(q_x)_B = \left(\frac{\Gamma}{4\pi}\right)\left(\frac{z}{x^2+z^2}\right)\left\{\left[\frac{(s/2-y)}{r(x,-y,z)}\right] + \left[\frac{(s/2+y)}{r(x,y,z)}\right]\right\},$$

plus three similar expressions for the components $(q_y)_T$, $(q_z)_B$, and $(q_z)_T$ respectively and $r(x, y, z) = (x^2 + (s/2 + y)^2 + z^2)^{1/2}$.

The horseshoe vortex is the simplest model for a wing of finite span. The velocity field at the bound vortex position is particularly of interest. Here, where $x = z = 0$, the bound vortex makes no contribution, and the trailing vortices produce a vertical velocity $(q_z)_T$ called "downwash":

$$q_z = \left(\frac{-\Gamma}{4\pi}\right)\left\{\left[\frac{1}{(s/2 + y)}\right] + \left[\frac{1}{(s/2 - y)}\right]\right\}. \tag{1.35}$$

If the circulation varies with y – the more general case – then the downwash at a point y of a span section dy located at η is

$$dq_z = \frac{d\Gamma(\eta)}{4\pi(y - \eta)}, \tag{1.36}$$

and the downwash of a wing with span s is

$$q_z(y) = \left(\frac{1}{4\pi}\right)\int_{-s/2}^{+s/2}\left(\frac{1}{y - \eta}\right)\left(\frac{d\Gamma}{d\eta}\right)d\eta. \tag{1.37}$$

It is worth noting at this point that lift L and circulation Γ are connected by the Kutta–Joukowsky expression for a wing section dy:

$$dL = \rho q \Gamma(y) dy, \tag{1.38}$$

where q is the induced velocity of the wing section.

The Γ distribution over the wing is a function of

(a) Wing geometry
(b) Angle of attack (angle between the chord and the incident fluid velocity vector)

The lifting-line model can be made more realistic by extending it to a nonzero chord (Fig. 1.14a) or by the use of a "vortex lattice" method in which cells consisting of vortex lines approximate the wing and the trailing vortices (Fig. 1.14b). Finally, the interaction of the long vortex wakes behind airplanes can be studied by a vortex tube model of the trailing vortices (see [48]) – Fig. 1.14c.

1.2.5.2 Vortex Rings

Vortex ring is the name given to a vortex tube which forms a closed loop.

Another curved vortex tube is the helical vortex. Figure 1.15 is a sketch of both of these vortices. They induce a velocity parallel to the ring axis and the helix axis, respectively, and travel *through space* at a constant speed. The vortex ring is the

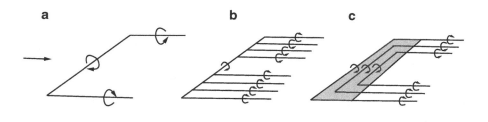

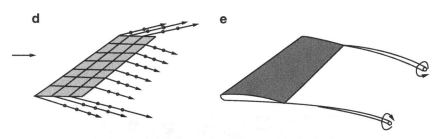

Fig. 1.14

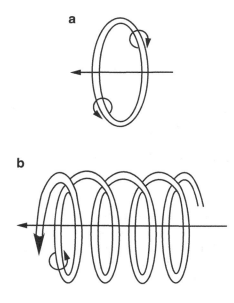

Fig. 1.15

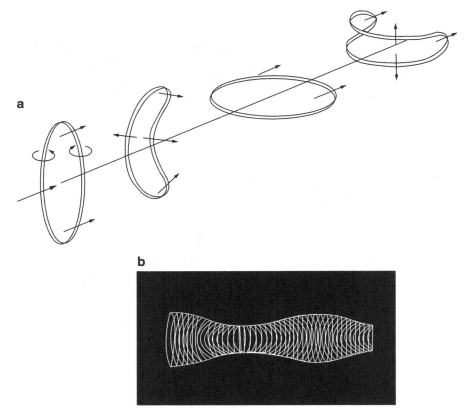

Fig. 1.16

more commonly observed and will be dealt with here. Smoke rings are perhaps the most frequently observed form and spectacular vortex rings are seen in explosions. Also impressive is the vortex ring formed in water. The ring is formed by an impulsive pressure applied to an orifice. By injecting a dye into the pressure chamber, a colored vortex ring emerges from the orifice and can be easily seen moving rapidly through the water chamber. It dissipates relatively slowly, as the surrounding fluid is viscous, and so the forward velocity decreases and the vortex eventually becomes unstable and breaks up.

Vortex rings can be rendered unstable by certain disturbances [46, 49] and others will result in oscillation. The oscillation of an elliptic vortex ring is sketched in Fig. 1.16. Oscillations, or pulsations, occur both in the plane of the ring and perpendicular to it. The motion is a result of repeated "catch-up" as the part of

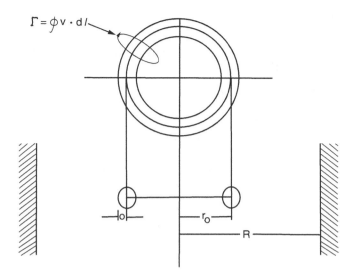

$\Gamma = \oint v \cdot d\,l$

Fig. 1.17

the ellipse with the greater curvature travels faster than the less curved part. In moving faster, the curvature diminishes and that part slows down and is passed out by the now more curved originally slower part. The cycle then repeats. This observation prompted Thompson (1867) to put forward the theory of vortex ring atoms, which described spectral lines in terms of the different modes of oscillation of vortex rings. Even though the theory was later discarded, it led to a great deal of analysis of vortex rings.

1.2.6 Translational Velocity of the Inviscid Vortex Ring

Consider a circular vortex filament, every element of which is rotating with angular velocity, ω, about the tangent to the circle of which the element forms a part. Figure 1.17 is a sketch of the cylindrical coordinate system and notation to be used. The z-axis passes through the center of the circle and is perpendicular to the plane of the circle. When the flow pattern is identical in any of the planes passing through the axis z, the flow system is axisymmetric and any point in the field is described by the coordinates(r, z). The equation of the streamlines is then

$$r(w\mathrm{d}r - u\mathrm{d}z) = 0, \tag{1.39}$$

where u is the axial component of fluid velocity and w is the radial component.
The equation of continuity is

$$\frac{\mathrm{d}(ru)}{\mathrm{d}r} + \frac{r\mathrm{d}w}{\mathrm{d}z} = 0. \tag{1.40}$$

In terms of the Stokes stream function ψ then

$$w = \frac{(1/r)\mathrm{d}\psi}{\mathrm{d}r}, \quad u = \frac{-(1/r)\mathrm{d}\psi}{\mathrm{d}z}. \tag{1.41}$$

The vorticity is

$$2\omega = \frac{\partial u}{\partial z} - \frac{\partial w}{\partial r}. \tag{1.42}$$

Substituting for u and w from (1.32), one obtains an equation for ψ and ω:

$$\frac{\partial^2 \psi}{\partial z^2} + \frac{\partial^2 \psi}{\partial r^2} - \left(\frac{1}{r}\right)\frac{\partial \psi}{\partial r} + 2r\omega = 0. \tag{1.43}$$

A circular vortex ring is supposed to comprise a large number of circular vortex filaments. At all points in the core of the ring (1.43) holds, while at exterior points, ψ satisfies the equation

$$\frac{\partial^2 \psi'}{\partial z^2} + \frac{\partial^2 \psi'}{\partial r^2} - \left(\frac{1}{r}\right)\frac{\partial \psi'}{\partial r} = 0. \tag{1.44}$$

Let $\psi = \chi r$, and then (1.43) and (1.44) become

$$\frac{\partial^2 \chi}{\partial z^2} + \frac{\partial^2 \chi}{\partial r^2} + \left(\frac{1}{r}\right)\frac{\partial \chi}{\partial r} - \frac{\chi}{r^2} + 2\omega = 0, \tag{1.45}$$

and

$$\frac{\partial^2 \chi'}{\partial z^2} + \frac{\partial^2 \chi'}{\partial r^2} + \left(\frac{1}{r}\right)\frac{\partial \chi'}{\partial r} - \frac{\chi'}{r^2} = 0. \tag{1.46}$$

Now $\Gamma = \pi a^2 \omega$ and by using elliptic integrals, defining a modulus as

$$k = \frac{2(rr_0)^{1/2}}{\left[z^2 + (r + r_0)^2\right]^{1/2}}, \tag{1.47}$$

and $k' = (1 - k^2)^{1/2} = a/2r_0$ at the surface of the ring, where r_0 is the radius of the ring and a is the core radius, the author has developed expressions for the radial and axial velocities, u and w, respectively [47]:

$$u = \left(\frac{\Gamma}{\pi}\right)\left\{\left(\frac{2r_0}{a}\right) - \frac{3a}{4r_0(L - 3/2)}\right\}\frac{z - z'}{2r_0 a}. \tag{1.48}$$

When $z = z'$, then $u = 0$ and the radius of the ring remains constant:

$$w = \frac{-\Gamma}{\pi}\left\{\left(\frac{2r_0}{a}\right) - \left(\frac{3a}{4r_0}\right)\left(\frac{L}{3/2}\right)\right\}\frac{2r_0(r - r_0) - a^2}{4r_0^2 a} + \frac{\Gamma}{2\pi r_0}(L - 2). \tag{1.49}$$

To obtain the velocity of the ring through the surrounding fluid one must put $r = r_0$ and then

$$w = \frac{\Gamma}{2\pi r_0}(L - 1) = \left(\frac{\Gamma}{2\pi r_0}\right)\left(\log\left(\frac{8r_0}{a}\right) - 1\right). \tag{1.50}$$

Thus, the ring will move forward in the direction of the cyclic motion through its aperture, with the constant velocity as given by (1.50). A simple calculation shows that the velocity produced by the vortex at the center of the ring is

$$\frac{\Gamma}{r_0} = \frac{\pi a^2 \omega}{r_0}.$$

Hence, an isolated vortex ring in an unbounded ideal fluid will move without change of size, in a direction perpendicular to its plane with a constant velocity.

1.2.7 Velocity of a Viscous Vortex Ring

In reality, vortex rings are viscous in nature. One can then define a Reynolds number (R_e) which indicates the importance of the fluid viscosity (v). R_e can be defined in several ways. In experimentation it is convenient to use parameters which are easy to measure and so

$$R_e = \frac{W_t D}{v}, \tag{1.51}$$

where D is the ring's diameter. Another one is based on the impulse and has been used in numerical simulation by Stananway et al. [51]:

$$R_e = \frac{[\sqrt{z(I/\rho)}]}{v t^{1/2}}. \tag{1.52}$$

Real vortex rings do not translate with constant velocity as expected from inviscid theory, but with a decaying velocity. Two useful asymptotic estimates for the velocity of a *viscous* vortex ring have been derived. One applies to vortex rings with thin cores ($a/R \to 0$) and is due to Saffman [50]. The other is due to Stanaway et al. [51] and applies to any viscous vortex ring at large times ($t \to \infty$). By assuming a Gaussian vorticity distribution in the core, Saffman derived an expression for the propagation speed (W_t) of a viscous vortex ring with a thin core:

$$W_t = (\Gamma(0)/4\pi R)\{\log 8R/(\sqrt{4vt}) - 0.588 + O(\sqrt{(vt/R^2)}\log(vt/R^2))\}. \quad (1.53)$$

1.2.8   Hydrodynamic Impulse and Vortex Ring Generation

Consider the fluid motion produced by impulsive forces. These are forces which act for very short times and so the acceleration terms $\partial u/\partial t$, etc. are much larger than the inertial terms $u\partial u/\partial x$, etc. The equations of motion for a perfect fluid now become

$$\frac{\partial u}{\partial t} = -\left(\frac{1}{\rho}\right)\frac{\partial p}{\partial x} + \frac{f_x}{\rho},$$

$$\frac{\partial v}{\partial t} = -\left(\frac{1}{\rho}\right)\frac{\partial p}{\partial y} + \frac{f_y}{\rho},$$

$$\frac{\partial w}{\partial t} = -\left(\frac{1}{\rho}\right)\frac{\partial p}{\partial z} + \frac{f_z}{\rho}, \quad (1.54)$$

where f_x, f_y, and f_z are the components of the impulsive force.

With the equation of continuity, there are then four linear equations in the four unknowns u, v, w, and p. The pressure can be eliminated from the equations by cross-differentiation. Now, the x component of vorticity is

$$\zeta_x = 2\omega_x = 2\left(\frac{\partial w}{\partial y} - \frac{\partial v}{\partial z}\right),$$

with similar expressions for the y and z components, and after some manipulation the following equation emerges

$$\frac{\partial \zeta_x}{\partial t} = \left(\frac{1}{\rho}\right)\left(\frac{\partial f_z}{\partial y} - \frac{\partial f_y}{\partial z}\right), \quad (1.55)$$

and two similar equations in ζ_y and ζ_z [47].

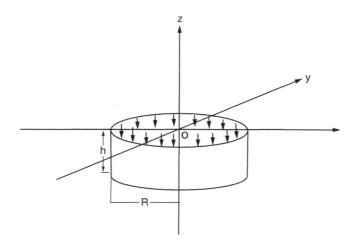

Fig. 1.18

Now consider the space lying between the planes $z = 0$ and $z = h$, bounded by a cylindrical surface of radius a, with its axis parallel to the z-axis (Fig. 1.18). Outside this region the external forces are taken to be zero, and inside the region they are directed in the negative direction. The force is constant across the region, but falls rapidly to zero at the cylindrical surface.

As f_z is the only force component acting in this case (1.55) reduce to

$$\frac{\partial \zeta_x}{\partial t} = \left(\frac{1}{\rho}\right)\frac{\partial f_z}{\partial y}, \quad \frac{\partial \zeta_y}{\partial t} = -\left(\frac{1}{\rho}\right)\frac{\partial f_z}{\partial x}, \quad \frac{\partial \zeta_z}{\partial t} = 0. \tag{1.56}$$

Since f_z varies only near the cylindrical surface, it is only in this region that vorticity is produced. If the motion starts from rest, and integrating (1.56) with respect to time, then

$$\zeta_x = \left(\frac{1}{\rho}\right)\frac{\partial I_z}{\partial y}, \quad \zeta_y = -\left(\frac{1}{\rho}\right)\frac{\partial I_z}{\partial x}, \quad \zeta_z = 0, \tag{1.57}$$

where I_z is the impulse (time integral) per unit volume of the force f_z. It is seen that the system of vortex lines generated in this way will consist of circles, with planes parallel to the xy plane and axes along the z direction. If the height h is much smaller than the circle radius R, then it is sufficient to combine these vortex lines into one single circular vortex lying in the xy plane and of radius R.

Figure 1.19 shows the section of the vortex in the xz plane, with circulation of fluid about it. The hatched region is that where f_z falls to zero.

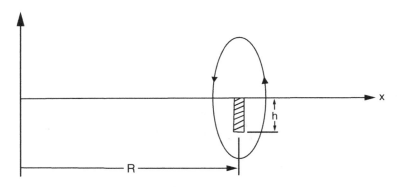

Fig. 1.19

The total strength Γ of the vortex is obtained by integrating ζ_y over this region. Therefore,

$$\Gamma = \iint dxdy\zeta_y = -\frac{1}{\rho}\int dz\int dx \frac{\partial I_z}{\partial x} = \frac{1}{\rho}\int I_z dz. \tag{1.58}$$

Since I_z is independent of z between the planes $z = 0$ and $z = h$,

$$\Gamma = \frac{I_z h}{\rho}. \tag{1.59}$$

When h is very small, $I_z h$ is the intensity per unit area of the impulse of the external forces (or the impulse of the pressure):

$$\therefore \Gamma = \frac{I}{\rho}. \tag{1.60}$$

A continuous force can be regarded as a sequence of impulses – each generating a vortex ring. These impulses move off from the region of formation with the fluid. In the limit of an infinite number of pulses, the rings merge into a continuous sheet of vorticity. This is illustrated in Fig. 1.20. The cylindrical surface becomes a vortex sheet. As shown previously, the strength Γ of a vortex sheet is determined by the difference of velocity on either side of the sheet and equals the circulation around a strip of the sheet having unit length in the direction of the cylinder axis. In the case in which the fluid is moving under a constant pressure P at a velocity V, the circulation generated in unit time is $V\Gamma$ and so

$$V\Gamma = \frac{P}{\rho}. \tag{1.61}$$

Fig. 1.20

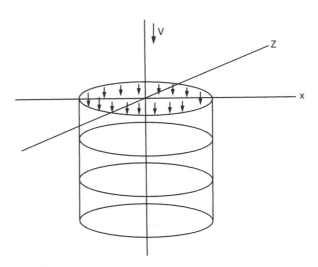

Vortex sheets are unstable and tend to roll up into a sequence of discrete vortices. This phenomenon will be dealt with later.

1.2.9 The Spectral Range of Vortex Size

The enormous range of vortex size has already been alluded to. At one end of the range, there are the quantized vortices in superfluid liquid helium. At the other end, there are the enormous rotational galactic systems. They also vary in morphology; rings, helices, spirals, swirls, etc. The following table [52] gives some idea of the great variety and is ordered in sequence of size:

Quantized liquid helium	10^{-8} cm
Small turbulent eddy	0.1 cm
Insect generated	0.1–10 cm
Dust whirls and whirlpools	1–10 m
Volcanic vortex rings and waterspouts	100–1,000 m
Convection clouds, hurricanes, and gulf stream	100–2,000 km
Ocean and atmospheric circulations and earth interior convection cells	2,000–5,000 km
Planetary atmospheres, Saturn rings, and sunspots	5,000–10^5 km
Interior star rotations	Varies with star size
Galaxies	Order of light years

Vortices can form in any of the known fluid states of matter – air, water, gas, plasma, and in molten solids. They can be generated by gradients in temperature, density, through frictional, electrodynamic, and gravitational forces. As mentioned earlier, in spite of this great variety in methods of generation, size, morphology, angular velocity, and kind of medium, they have common characteristics and structures. In approaching these aspects of the vortex phenomenon, it will be necessary to utilize analytical and computational mathematics, but only so far as they can be used to deal with the physics of the vortex.

1.3   Forces (Lift, Drag, Thrust, and Torque) on Moving Submerged Bodies Due to Vortex Formation

The conservation laws and the constitutive equations form the two basic axioms of fluid dynamics [16]. Conservation laws exist for all forms of matter, all forms of energy, momentum, and angular momentum. The constitutive equations concern the properties of matter, whether elastic, rigid, liquid, plastic, or a gas.

A body in a flowing fluid will experience two main forces on it which are perpendicular to one another – the lift which is perpendicular to the flow direction and the drag which is parallel to the flow. If not in space, there will also be the gravitational force in a direction opposite to the lift – see Fig. 1.21.

If the resultant force on the body does not act through the center of gravity of the body, there will be a torque on the body which will cause the body to rotate.

1.3.1   Aerodynamic Lift and Drag

In the nineteenth century, the challenge of solving the problem of flight – both powered and unpowered (gliding) – received increasing attention [53, 54]. Considering a wing or airfoil in a flowing inviscid fluid, and using Bernoulli's equation, the pressure on the surface of the wing can be determined from the velocity distribution adjacent to it. In inviscid flow, there is no boundary layer and integration of the pressure around the wing yielded the result that the wing experiences neither drag nor lift! This contradicted all experience – in both gliding and observing birds. However, in 1894, an Englishman named Lanchester found the solution.

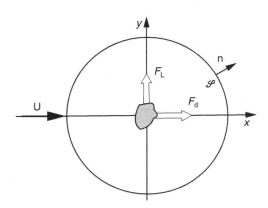

Fig. 1.21

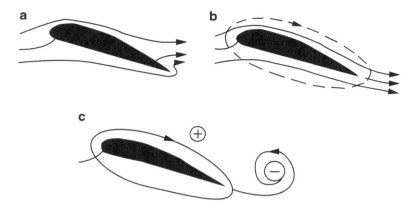

Fig. 1.22

He suggested that a vortex acting on the wing produced the lift. This is best understood in terms of the Magnus effect, named after Heinrich Magnus (1802–1870), a German physicist. Consider the ideal flow about a cylinder as illustrated in Fig. 1.22a. There is no boundary or separation and symmetry of the streamlines results in equal pressures fore and aft. The cylinder in ideal cross-flow experiences no drag. But consider another ideal flow of concentric streamlines – Fig. 1.22b. Superposing these two flows produces the streamlines shown in Fig. 1.22c. Now asymmetry appears the flow – streamlines of (a) and (b) flow in the same direction on top, and in opposite directions on the bottom of the cylinder. Using Bernoulli's theorem along with the conservation of volume in incompressible flow shows that there is now low pressure at the top and high pressure below the cylinder. A net lift force acts on the cylinder at right angles to the flow direction. Lanchester's hypothesis was later formulated in mathematical terms by the German mathematician Wilhelm Kutta and the Russian scientist Nikolai Joukowsky, acting independently.

They put forward the theory of lift in mathematical terms and quantified the strength of the circulation about the wing (Fig. 1.23). One notes that the rear stagnation point moves to the rear of the wing – the trailing edge as it is called. Very large velocity differences can theoretically arise. It was Kutta who found a way of correcting this unrealistic mathematical prediction. In order to have a finite velocity at the trailing edge under some given flight conditions, the strength of the vortex must be such as to ensure smooth flow at that location (see Fig. 1.24b). This criterion for smooth flow is known as the Kutta condition.

Although lift is obtained with this model, there is no drag on the body. Lanchester took care of this, and improved his theory, by assuming a wing of finite length. At the ends of the wing, two vortices are formed as a result of the unequal

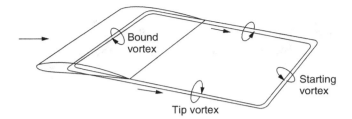

Fig. 1.23

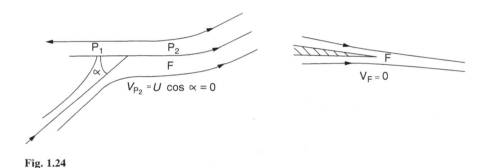

Fig. 1.24

pressures above and below the wing. It is these "tip vortices" at the ends which produce drag. Together with the bound vortex they form what is called a "horseshoe vortex." As per Helmholtz's first theorem, this vortex must be closed and the "starting vortex," which is formed when the plate, or wing, first moves, closes the vortex ring – see Fig. 1.25. Independent of Lanchester, Prandtl later developed the hypothesis of wing circulation and the tip vortex [55]. From these beginnings and the development of boundary-layer theory, the prestigious Gottingen School formed around Prandtl and was a key center in the development of aerodynamic theory between World Wars 1 and 2.

The flat plate with a nonzero angle of attack[1] does not produce the greatest lift. In fact, the optimal shape of a wing profile is multi-factorial. Aircraft were already highly developed by the time realistic mathematical theories of lift were developed [56]. In fact, the achievement of accurate design of efficient wings has had to await the opening of the new field of computational fluid dynamics.

---

[1] Defined as the angle between the chord line and the flight direction.

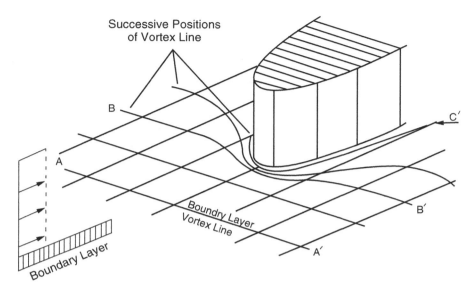

Successive Positions
of Vortex Line

B

A

Boundary Layer
Vortex Line

Boundary Layer

C′

B′

A′

Fig. 1.25

1.3.2   Analytical Derivation of Lift and Drag on a Cylinder

In terms of circulation and vortices it is possible, and instructive, to derive elegant
solutions to derive expressions for the lift and drag on a cylinder in an ideal fluid.

It is found that the combination of a uniform flow, a doublet and a vortex, can
represent the flow past a cylinder with circulation [48, p. 182]. The stream function
and velocity potential are, respectively,

$$\psi = U_\infty \left(r - \frac{R^2}{r^2} \right) \sin\theta + \left(\frac{\Gamma}{2\pi} \right) \ln r, \tag{1.62}$$

$$\phi = U_\infty \left(r + \frac{R^2}{r^2} \right) \cos\theta - \left(\frac{\Gamma}{2\pi} \right) \ln r, \tag{1.63}$$

where r, θ are cylindrical coordinates with the origin at the center of the cylinder, R
is the cylinder radius, and Γ is the strength (circulation) of the vortex. The velocity
components are, therefore,

$$v_\theta = -\frac{\partial \psi}{\partial r} = -U_\infty \left(1 + \frac{R^2}{r^2} \right) \sin\theta - \frac{\Gamma}{2\pi r}, \tag{1.64}$$

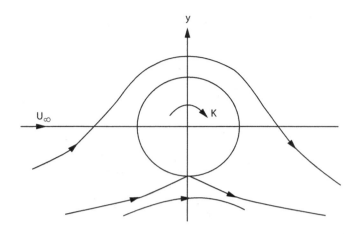

Fig. 1.26

$$v_r = \left(\frac{1}{r}\right)\frac{\partial \psi}{\partial \theta} = U_\infty \left(1 - \frac{R^2}{r^2}\right)\cos\theta. \tag{1.65}$$

The radial velocity component, in fact, is not changed by the circulation. At the surface of the cylinder ($r = R$),

$$v_\theta = -2U_\infty \sin\theta - \frac{\Gamma}{2\pi R}. \tag{1.66}$$

At the stagnation points, $v_\theta = 0$ and so

$$\sin(-\theta_S) = \frac{\Gamma}{4\pi U_\infty} = \sin(\pi + \theta_S), \tag{1.67}$$

where $\Gamma/4\pi R U_\infty$ is less than or equal to zero.

When $\Gamma = 4\pi R U_\infty$, the two stagnation points coincide at $r=R$ and $\theta=-\pi/2$ (see Fig. 1.26). When Γ is greater than this, the stagnation points move away from the cylinder surface. This can be shown as follows. The condition for a stagnation point away from the cylinder is that *both* velocity components be zero. This can only be satisfied if $\theta = 3\pi/2$ (as Γ cannot be negative), and using this value the values of r at which the stagnation points occur are given by

$$r = \frac{\Gamma}{4\pi U_\infty} + \left(\frac{1}{2}\right)\left[\left(\frac{\Gamma}{2\pi U_\infty}\right)^2 - 4R^2\right]^{1/2}. \tag{1.68}$$

Fig. 1.27

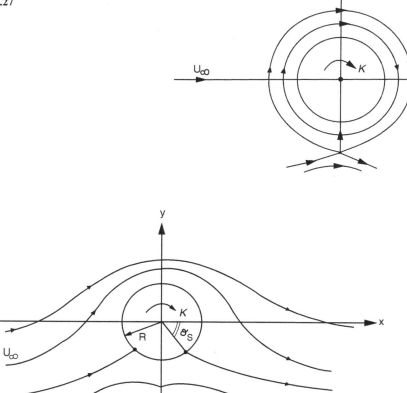

Fig. 1.28

Moreover, the points must lie along the negative y-axis (Fig. 1.27).

For smaller values of Γ the stagnation points lie on the cylinder, and the position for any R, Γ, and U_∞ is obtained by solving (1.67) for θ_S (see Fig. 1.28).

At large distances from the cylinder, the velocity and the pressure p_∞ in the approaching flow are uniform. The equation for the pressure p_S at any point on the surface of the cylinder is obtained from Bernoulli's equation:

$$p_S = p_\infty + \left(\frac{1}{2}\right)\rho U_\infty^2 - \left(\frac{1}{2}\right)\rho U_S^2. \qquad (1.69)$$

From (1.65) the tangential velocity at the cylinder surface, which is U_S, is

$$(v_\theta)_S = U_S = -2U_\infty \sin\theta - \Gamma/2\pi R.$$

Using this in (1.69), one obtains

$$p_S = p_\infty + \left(\frac{1}{2}\right)\rho\left[U_\infty^2 - \left(2U_\infty \sin\theta + \frac{\Gamma}{2\pi R}\right)^2\right]. \qquad (1.70)$$

This pressure is normal to the cylinder surface and results in an elemental force $p_S R d\theta$ on an elementary area $R d\theta$ per unit length of the cylinder. The total lift and drag forces are, thus,

$$L = \int^{2\pi} - p_S R \sin\theta d\theta, \qquad (1.71)$$

$$D = \int^{2\pi} - p_S R \cos\theta d\theta. \qquad (1.72)$$

Because of the symmetry of the flow pattern about a vertical axis through the cylinder center, the integral for D is zero, and so irrotational flow theory again predicts zero drag in this case.

By substituting for p_S from (1.70), the lift L per unit length can be determined:

$$L = \left(\frac{\rho U_\infty \Gamma}{\pi}\right)\int^{2\pi} \sin^2\theta d\theta = \rho U_\infty \Gamma. \qquad (1.73)$$

This is known as the Kutta–Joukowski theorem. The lift is thus directly proportional to

(a) The fluid density
(b) The stream velocity
(c) The circulation

By use of the mathematics of complex variables and conformal mapping, it is possible to transform a two-dimensional irrotational flow about a circular cylinder to a flow about various airfoil shapes. The lift force produced by circulation of fluid about the cylinder can also be produced by rotation of the cylinder in a fluid stream. This phenomenon was first observed by the German scientist Magnus in 1852. It is usually referred to as the Magnus effect. A well-known consequence of the effect is the curved trajectory of a spinning baseball.

1.3.3   Axial Thrust, Torque, and Helical Vortices

In a world where transportation is of supreme importance, axial thrust is the most efficient force. Excluding the reactionary force involved in rocket propulsion, the

Fig. 1.29

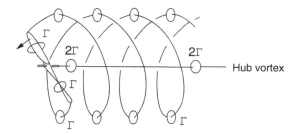

Hub vortex

most important devices for producing axial thrust are propellers, turbines, and water screws. Vortex formation is again basic to their kinematics, with helical, hub, and bound vortices, all playing a part.

For an airplane or ship propeller, the forward thrust is generally horizontally directed, while for a helicopter the thrust is mainly vertical to counteract gravity. For the sake of simplicity, the vortex system of the propeller can be represented by a vortex array consisting of hub, bound, and helical vortices (Fig. 1.29).

The differential force acting at any radius r is, by the Kutta–Joukowsky equation (1.73) applied locally at the bound vortex,

$$dF = \rho U \Gamma(r). \tag{1.74}$$

This force is perpendicular to the resultant velocity U which includes the induced velocities and the relative axial and tangential components $-U_{ax}$ and $-r\omega$ [57]. The thrust is generated by the net tangential velocity and the torque by the net axial component.

The axial and tangential components are then

$$dF_{ax} = \rho U \Gamma \cos\beta_i, \tag{1.75}$$

$$dF_t = -\rho U \Gamma \sin\beta_i. \tag{1.76}$$

The total thrust and torque for a Z-bladed propeller are derived by integrating, summing, and changing sign on the force equations to define the thrust and torque in accordance with current industrial practice.

This results in the following equations:

$$T = \rho z \int_{Rh}^{R} \left[U\Gamma\cos\beta_i - \left(\frac{1}{2}\right)c(r)C_D \sin\beta_i U^2 \right] dr, \tag{1.77}$$

$$Q = \rho z \int_{Rh}^{R} \left[U\Gamma \sin\beta_i + \left(\frac{1}{2}\right)c(r)C_D\cos\beta_i U^2 \right] dr, \tag{1.78}$$

where $c(r)$ is the blade element of chord[2], C_D is the viscous drag coefficient, R is the propeller radius, and R_h is the hub radius.

The full development of these equations and how they are used in the design and analysis of propellers are given in [57]. It should be noted that the z value for the propeller is as per ITTC notation and not to be confused with, although obviously related to, the Z used as the argument in the Legendre polynomials which crop up in propeller hydrodynamic analyses and always defined as

$$Z = \frac{\left(x - x^|\right)^2 + r^2 + r^{|2}}{2rr^|}.$$ (1.79)

For a constant value of Z this defines a torus in which the locus of centers is given by $R_c = Zr^|$ and the radius is given by $R_r = r^|\sqrt{(Z^2 - 1)}$.

1.4   Some Other Kinds of Simple Vortices

1.4.1   Intake and Inlet Vortices

These vortices occur when an intake – usually indicating some kind of pipe opening – or inlet – usually indicating an aircraft gas turbine engine – operates near the ground at near-static conditions. Blanchette [58] has designed an experiment which demonstrates the generation of a vortex stretching from a solid surface into a pipe intake. One requires a fan to generate an airflow, a vacuum to suck air into the intake, and a table. When there is no table nearby and the air being sucked into the intake is not disturbed, an axisymmetric "sink flow" is produced at the intake tube of the vacuum cleaner – see Fig. 1.30a. When the tabletop is placed near the intake, the symmetry is destroyed and the axis of symmetry – the dashed line in Fig. 1.30a – is now bent toward the table – see Fig. 1.30b. When the table is removed and the fan's air stream is arranged at an angle to the axis of symmetry, an intake vortex is created with the axis of symmetry also being the rotation axis – see Fig. 1.30c. When the table is again placed near the intake, the rotation axis bends toward the table – Fig. 1.30d. The vortex can be visualized by replacing the table with water surface. The low pressure at the vortex axis causes the free surface of the water to bulge out, and the rising air entrains water droplets and creates a miniature waterspout.

It has been proposed that the cause of the intake, or inlet, vortex is the stretching of ambient vorticity and its amplification as it is drawn into the intake (or engine inlet). However, because vortex lines cannot end in a fluid, two vortices must enter the inlet, with equal and opposite circulation – see Fig. 1.31. Only one vortex is

---

[2] The chord is defined as the distance from the leading edge to the trailing edge.

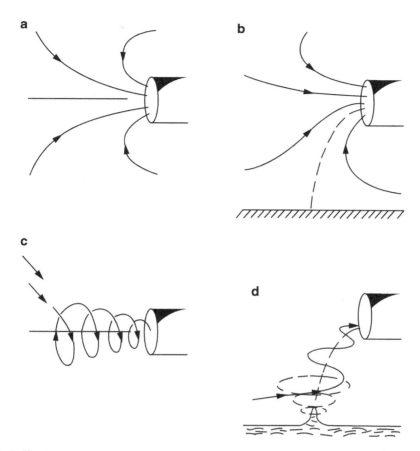

Fig. 1.30

observed in practice and so there must be another vortex kinematic mechanism at work. By experiment and computation, it has been shown [59, 60] that ambient vortex lines perpendicular to the primary flow and parallel to the ground plane (these lines are horizontal in the case of an aircraft engine inlet) can evolve into a pair of counter-rotating vortices, of approximate equal strength, on the inlet face. However, for an ambient vorticity field consisting of vertical vortex lines, the situation is different. The vortex lines as they are convected toward the inlet by the primary flow evolve into a configuration in which the upper parts of the vortex lines are spread out over the upper part of the inlet, while the lower parts concentrate about the stagnation streamline connected with the stagnation point on the

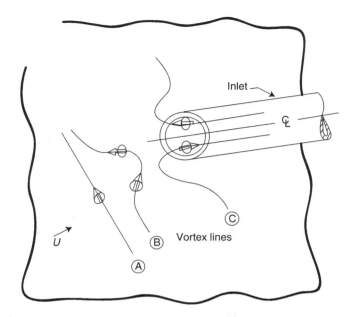

Fig. 1.31

ground plane. Figure 1.32 is based on particle track computations and shows the concentration of vortex lines in the lower part of the inlet [61]. This concentration delineates the region of increased circulation per unit area associated with the observed inlet vortex.

Inlet vortex formation in irrotational flow: the mechanism for inlet vortex formation described above involved stretching of ambient vertical vortex lines. A second mechanism occurs when there is a cross-wind and occurs even when the upstream velocity field is irrotational (no ambient vorticity). One vortex extends from the ground plane to the lower part of the inlet. A second *trailing* vortex (larger than the inlet) is observed to leave from the downstream side of the inlet lip, as shown in Fig. 1.33 [60]. In the direction of the inlet the rotation of the inlet vortex is clockwise, and that of the trailing vortex is clockwise when viewed from upstream.

The formation of the trailing vortex correlates with the variation of circulation along the inlet. Several diameters downstream from the lip of the inlet the velocity across the inlet will be close to the free stream velocity, U_∞. The circulation around the contour C_1 at this location – see Fig. 1.34 – will be much less than that around the contour C_2 at the inlet lip [61]. The difference in circulation around

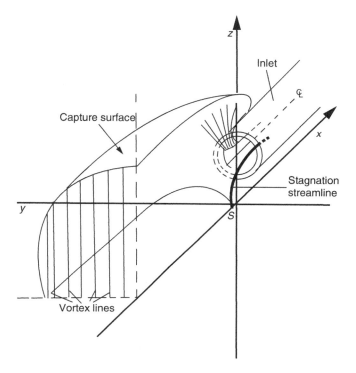

Fig. 1.32

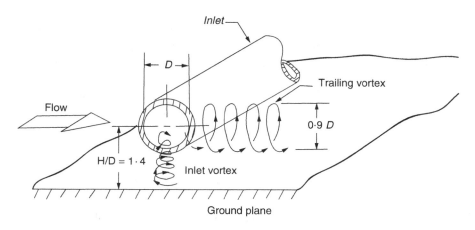

Fig. 1.33

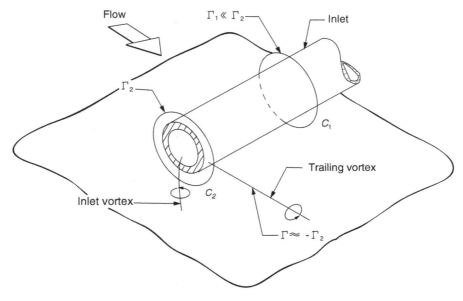

Fig. 1.34

these two contours implies trailing vorticity between the two locations as indicated in Fig. 1.34, and the circulation of the trailing vortex is approximately equal and opposite to that of the inlet vortex.

The pressure field associated with the inlet vortex system results in an asymmetric flow pattern around the front part of the inlet. The separation line extends from the 12 o'clock position at a downstream location to a 4 o'clock position at the inlet lip – see Fig. 1.35 [60]. The skew in the separation line is due to the increase in circulation around the inlet as one moves toward the inlet lip. Vortex lines can leave or enter the inlet surface viscous layers only along this separation line. Although there is some analogy between the inlet vortex system and that of a finite wing, the way in which the vortex lines enter or leave the inlet surface is basically different from the situation for a finite wing.

1.4.2   Clearance Vortices (Tip Clearance Flows)

Tip clearance flows occur commonly in turbomachinery – for example, between the rotor blades and the outer casing in compressors and turbines. The pressure

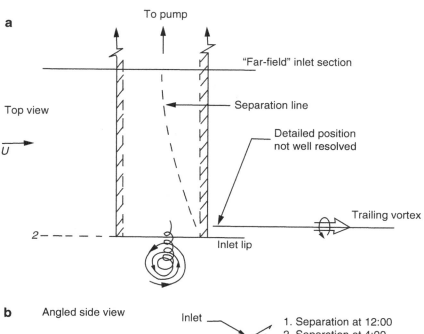

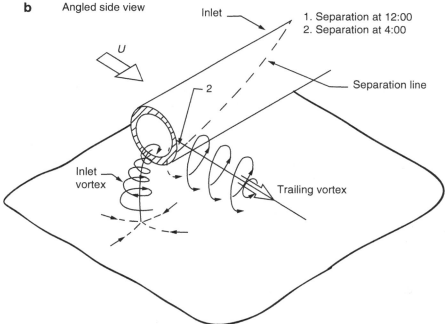

Fig. 1.35

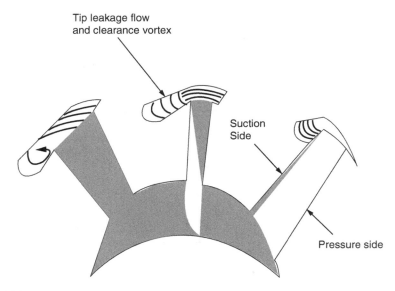

Fig. 1.36

difference across the blade results in a jet flow through the clearing, which rolls up
to form a clearance vortex. This is illustrated in Fig. 1.36.

The three-dimensional steady clearance flow can be approximated by a two-
dimensional unsteady process. The generation of the tip clearance flow can then be
regarded as a sequence of events in successive cross-flow planes perpendicular to
the blade camber [62]. This is illustrated in Fig. 1.37. A, B, C, and D are the cross-
flow planes, with location (a) at the leading edge and location (d) at the trailing
edge. As one moves through the planes (representing the blade passage), the vortex
sheet in the clearance region rolls up. Time, t, in the lower half of the figure, is
related to streamwise location, s, in the upper half by the relation, $t = s/V(s)$, where
$V(s)$ is the velocity of the moving reference frame.

Similarity analysis: in developing a similarity variable, viscous effects and
compressibility may be neglected [65, p. 491]. Figure 1.38 shows the blade and
flow domain at an arbitrary location. A useful quantity is the centroid of the rolled-
up part of the vortex sheet. There are four physical variables which can characterize
this situation and from which a dimensionless variable can be constructed: tip
clearance τ, pressure difference across the blade ΔP, density ρ, and time t. Nondi-
mensional time can be defined as

$$t* = (t/\rho)\sqrt{(\Delta P/\rho)}. \tag{1.80}$$

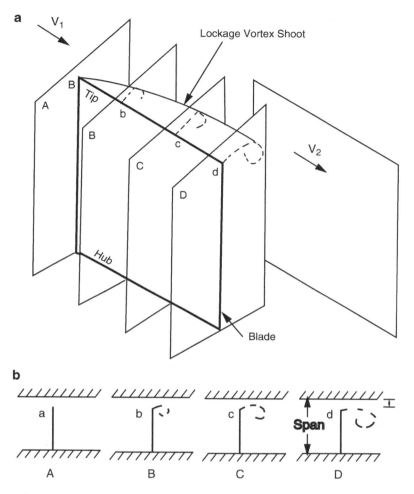

Fig. 1.37

Two tip clearance flows will be similar if they have the same t^* and some parameters will be functions of t^* only:

$$\text{Vortex center coordinates}: y_C^* = \frac{y_C}{\tau}, \quad z_C^* = \frac{z_C}{\tau},$$

$$\text{Vortex circulation (non - dim.)}: \Gamma* = \Gamma/(\tau\sqrt{\Delta P/\rho}),$$

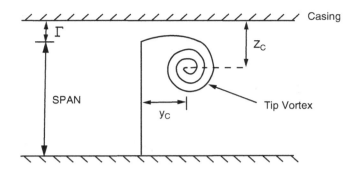

Fig. 1.38

Cross - flow velocity components : $v* = v/(\sqrt{\Delta P/\rho})$, $w* = w/(\sqrt{\Delta P/\rho})$.

The pressure difference across the blade does vary along the span, but experimental data show that evaluation of the mean loading at mid-radius is accurate enough. Equation (1.80) can therefore be written in terms of flow angles at the mean radius using the commonly used expression for ideal pressure rise [64]:

$$t* = (x/\tau)\sqrt{[g(\tan\beta_1 - \tan\beta_2)/c\cos\beta_m]}, \qquad (1.81)$$

where g is the blade spacing, c is the chord length, β_1 and β_2 are the inlet and outlet angles, respectively, β_m implies the mean velocity direction, and x is the axial coordinate.

Velocity and vorticity fields for the two-dimensional flow can be computed numerically [62] and are summarized in [63]. It is found that the generalized tip vortex trajectory is nearly a straight line and can be represented by the equation

$$y_c^* = 0.46t * . \qquad (1.82)$$

For further details, see [65, p. 494].

The radial motion of the tip clearance vortex can be explained by vortex dynamics and although this is somewhat invalidated because of turbulent diffusion near the wall, a useful approximation to some of the important flow features may be obtained. The trailing vortex can be regarded as a number of elementary line vortices moving in one another's velocity fields. For the flow bounded by the endwall plane, the flow is represented by a vortex system and its images so that each elementary vortex is a half of a vortex pair. The Kelvin impulse of a vortex pair is given by [65]

$$P_i = \rho \Gamma_i d_i, \tag{1.83}$$

where Γ_i is the circulation of the vortex and d_i is the distance between the centers of the vortex pair. The impulse of the N vortex pairs forming the trailing vortex is then

$$P = \Sigma_{i=1}^{N} \rho \Gamma_i d_i. \tag{1.84}$$

Downstream of the blades, there is no force on the fluid and the total Kelvin impulse is constant. Moreover, no vorticity is shed and the total circulation is also conserved:

$$\Gamma = \Sigma_{i=1}^{n} \Gamma_i = \text{const.} \tag{1.85}$$

Combining (1.84) and (1.85) gives

$$\frac{\Sigma_{i=1}^{n} \Gamma_i d_i}{\Sigma_{i=1}^{N} \Gamma_i} = \text{const.} = 2z_C. \tag{1.86}$$

This states that the radial position of the centroid of vorticity, z_C, which is effectively the center of the vortex core, stays at a fixed distance from the wall. Experimental data and three-dimensional computations confirm this [62].

1.4.3   Multiple Point Vortices and Their Motion

When two or more vortices are generated simultaneously, they interact and undergo characteristic motions. The case of two vortices will be considered and analyzed first. Two nearby vortices interact so that their rotating axes migrate. The velocity of the first vortex determines the velocity of the center of the second vortex, and the velocity of the second vortex determines the motion of the first. The resulting velocity is the superposition of velocities of the two vortices and so the vortex system itself has a translational velocity.

The projection of the line vortex in a perpendicular plane is a point and the streamlines are concentric circles about this point vortex. Denoting this plane as the xy plane and with the line vortex intersecting the plane at the point (ξ, η), the component velocities are

$$v_x = \frac{-\Gamma}{2\pi} \left[\frac{y - \eta}{r^2} \right], \quad v_y = \frac{-\Gamma}{2\pi} \left[\frac{x - \xi}{r^2} \right], \tag{1.87}$$

where $r^2 = (x - \xi)^2 + (y - \eta)^2$.

Due to the symmetry of the motion around a point (or line) vortex, the vortex will not translate. Induced motion occurs when

(a) The vortex line is curved.
(b) Two (or more) straight line vortices are near one another.

Consider two parallel rectilinear vortex filaments. From (1.87) the complex velocity can be constructed:

$$v_x - iv_y = \frac{-\Gamma}{2\pi} \left\{ \frac{(y - \eta) + i(x - \xi)}{r^2} \right\} = \frac{\Gamma}{2\pi i} \left[\frac{z* - z_0*}{(z - z_0)(z* - z_0*)} \right], \qquad (1.88)$$

where $z_0 = \xi + i\eta$, $z* = x - iy$, $z_0* = \xi - i\eta$,

The complex velocity equals the derivative of the complex potential ψ with respect to z:

$$v_x - iv_y = \frac{dw}{dz}, \qquad (1.89)$$

so that

$$\psi = \frac{\Gamma}{2\pi i} \log(z - z_0). \qquad (1.90)$$

The fluid motion takes place in a plane perpendicular to the two parallel filaments and this will be taken as the z plane. Suppose the vortex strengths are Γ_1 and Γ_2, respectively. The complex potential for the pair is then

$$\psi(z) = \frac{\Gamma_1}{2\pi i} \log (z - z_1) + \frac{\Gamma_2}{2\pi i} \log(z - z_2),$$

and the complex velocity is

$$v_x - iv_y = \frac{dw}{dz} = \frac{\Gamma_1}{2\pi i} \left[\frac{1}{z - z_1} + \left(\frac{\Gamma_2}{2\pi i} \right) \frac{1}{z - z_2} \right]. \qquad (1.91)$$

Now, $v_x - iv_y = dz*/dt$ and so

$$\frac{dz*}{dt} = \frac{\Gamma_1}{2\pi i} (z - z_1) + \frac{\Gamma_2}{2\pi i} (z - z_2). \qquad (1.92)$$

The vortex at the point z_1 moves solely under the influence of the other vortex at z_2. The first vortex will move in a circle about the second one and the second in a circle about the first. The intervortex distance remains constant during the motion.

This motion will now be proved. To obtain the velocity of the *first* vortex, the first term in (1.92) must be omitted and z replaced by z_1. Then,

$$\frac{dz_1^*}{dt} = \frac{\Gamma_2}{2\pi i}(z_1 - z_2). \tag{1.93}$$

Similarly,

$$\frac{dz_2^*}{dt} = \frac{\Gamma_1}{2\pi i}(z_2 - z_1). \tag{1.94}$$

Separating these equations into real and imaginary parts and after some manipulation [48, p. 210] it can be shown that

$$\frac{\Gamma_1 x_1 + \Gamma_2 x_2}{\Gamma_1 + \Gamma_2} = \text{constant}, \tag{1.94a}$$

$$\frac{\Gamma_1 y_1 + \Gamma_2 y_2}{\Gamma_1 + \Gamma_2} = \text{constant}. \tag{1.94b}$$

These are called "integrals of motion of the centroid" of the system of two vortices and they give the values of x_c and y_c, respectively, the centroid of the two vortices, which remains fixed at all times during the motion. The equation of the circle traced out by the two vortices is given by [48, p. 212]

$$(x_1 - x_2)^2 + (y_1 - y_2)^2 = \text{constant}. \tag{1.95}$$

Thus, the two vortices rotate about the centroid with constant distance between them – which was to be proved.

When $\Gamma_1 = -\Gamma_2$ (the vortices have equal and opposite rotation), the centroid will lie at infinity since $\Gamma_1 + \Gamma_2 = 0$. It will now be shown that these vortices will move in translation with constant velocity, perpendicular to the straight line joining them (Fig. 1.39). Suppose that initially the two vortices are on the Ox axis separated by a distance R. From the equations of motion,

$$\frac{dz_1*}{dt} = \frac{dz_2*}{dt} = \frac{-\Gamma_1}{2\pi i}(z_1 - z_2),$$

$$z_2* - z_1* = \text{constant} = R,$$

$$z_2 - z_1 = \frac{1}{R},$$

therefore,

$$\frac{dz_1*}{dt} = \frac{dz_2*}{dt} = \frac{\Gamma_1}{2\pi i R} = -\frac{\Gamma_1 i}{2\pi R}.$$

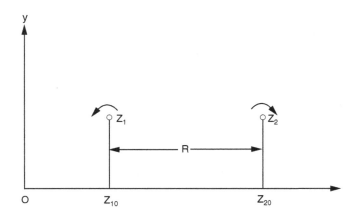

Fig. 1.39

Separating the real and imaginary parts, one obtains

$$v_{1x} = v_{2x} = 0, \quad v_{1y} = v_{2y} = \frac{\Gamma_1}{2\pi R}. \tag{1.96}$$

Thus, the two vortices move parallel to the Oy axis.

This case relates to the situation in which a vortex lies close to a wall. There is an image vortex of equal but opposite rotation in the wall, and so the vortex will move along the wall. If the vortex is in a corner, then there are three image vortices as shown in Fig. 1.40. There are two negative vortices at the points $(-x, y)$ and $(x, -y)$ and a positive vortex at the point $(-x, -y)$. Since the vortex at $P(x, y)$ cannot induce any self-motion, its motion is due solely to that produced by the combined effect of its images. Therefore,

$$\frac{dx}{dt} = \frac{\Gamma}{2\pi y} - \frac{\Gamma y}{2\pi(x^2 + y^2)} = \frac{\Gamma x^2}{2\pi y(x^2 + y^2)}, \tag{1.97a}$$

$$\frac{dy}{dt} = -\frac{\Gamma}{2\pi x} + \frac{\Gamma x}{2\pi(x^2 + y^2)} = -\frac{\Gamma y^2}{2\pi x(x^2 + y^2)}. \tag{1.97b}$$

Thus,

$$\frac{dx/dt}{x^3} + \frac{dy/dt}{y^3} = 0 \quad \text{or} \quad \frac{1}{x^2} + \frac{1}{y^2} = \frac{1}{a^2},$$

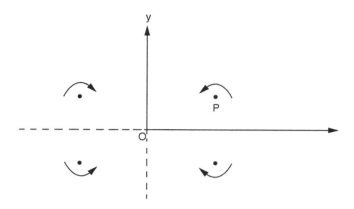

Fig. 1.40

where a is a constant.

Therefore, $(x^2 + y^2)/x^2y^2 = 1/a^2 = r^2/x^2y^2$, and so

$$\frac{r}{xy} = \frac{1}{a} = \frac{r}{r^2\cos\theta\sin\theta},$$

and finally

$$r\sin2\theta = 2a. \tag{1.98}$$

This is the equation of the curve described by the vortex. Moreover, since $(x\mathrm{d}y/\mathrm{d}t) - (y\mathrm{d}x/\mathrm{d}t) = -(1/2)\Gamma$, the vortex executes the curve that a particle would describe if repelled from the origin with a force equal to $3\Gamma^2/16\pi^2 r^3$.

An example of such a vortex system is the motion of two vortices behind an oar or at the wing tips of an airplane – see Fig. 1.41. These vortices persist for several minutes over the airfield after take-off and landing operations of large aircraft. They cause downdrafts of several meters per second and can endanger small aircraft which cross their paths. Figure 1.42 illustrates the motion of two point vortices with the same direction of rotation and equal and unequal strengths – in both cases, the vortices rotate about a center.

Vortex interaction on a large and spectacular scale occurs when two tropical whirlwinds meet. As these atmospheric vortices are nearly always cyclonic, they will have the same rotation and will rotate around one another, and form a major component of tropical storms. This is commonly called the "Fujihara effect" and has been observed [66]. More than two vortices: the motion of a straight chain of point vortices, equally strong and rotating in the same direction – see Fig. 1.43.

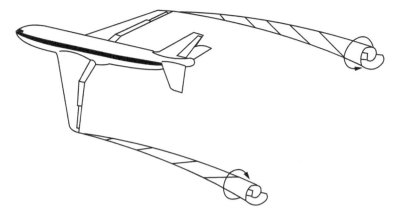

Fig. 1.41

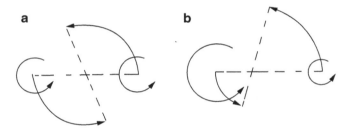

Fig. 1.42

This chain approximates a "discontinuity line" across which there is a jump in velocity. The line is an abstraction in which the vorticity would become infinite – it does not exist. It is useful for illustrative purposes. A small deviation in the linear array of point vortices would destroy the array and lead to instability. The chain would then roll up into a collection of larger vortices [67]. The model is not realistic as the roll-up process is a function of both the initial disturbance and the number of vortices per wavelength of the disturbance. Moreover, viscosity is neglected. A more realistic model is achieved if the vortex chain is replaced by a vortex band comprising a multitude of point vortices (Fig. 1.44; [68]). The motion of two vortex clusters, consisting of large numbers of point vortices, has also been simulated

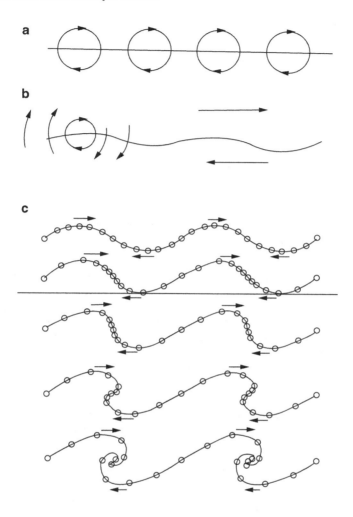

Fig. 1.43

using CFD [68]. When the two clusters are far enough from each other, they rotate around one another just as the two point vortices do. However, in this case, the circular areas deform and oscillations occur on the cluster surfaces (Fig. 1.45a). In contrast, when the two vortices are very close, they merge (Fig. 1.45b). In between these two cases a critical distance exists at which the vortex clusters alternately approach one another and move apart. In this process, point vortices are exchanged (Fig. 1.45c).

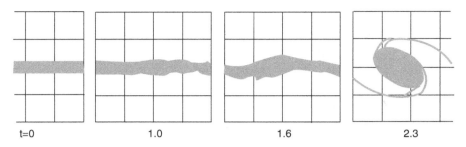

t=0   1.0   1.6   2.3

Fig. 1.44

1.4.4 Quantum Vortices and the Hydrodynamics of Superfluids

The understanding of liquid helium and its unusual physical properties is greatly dependent on experimental and theoretical studies of vortices [39, 42, 69–73]. The ^{4}He isotope was first liquefied in 1908. The liquid form is commonly designated as helium I or II, with ^{4}He II being the superfluid and superconducting form. For example, He II can flow through extremely fine capillaries with apparently no energy loss due to viscous friction – this property is lost when the speed of the liquid exceeds a certain critical velocity [74, 75].

Two-fluid model of He II: the "super" or peculiar properties of He II can be largely accounted for on the basis of phenomenological two-fluid theory [37, 38]. A most striking characteristic of liquid helium is that, as far as is known, it exists in a liquid state down to the absolute zero temperature. This is due to two factors:

(a) The van der Waals forces in helium are weak.
(b) The zero-point energy, due to the light mass, is large.

The atoms in the liquid are relatively far apart and considerable pressure is required (about 26 atm at 0 K) to force them into the solid state. Thus, the characteristics of liquid helium are governed by quantum laws and it is called a quantum liquid. The "λ-point" is the temperature at which the specific heat versus temperature undergoes a discontinuity. At temperatures above this point, helium is designated as He I and below it as He II. London in 1954 [39] suggested that below the λ-point helium is a quantum fluid whose essential feature is the macroscopic occupation of a single quantum state. This implies long-range order in He II and is called *Bose condensation*. The "condensed" units are held together by exchange forces and hence the whole condensate moves as a whole. In the ground state $|0 > a$, the macroscopic number of particles n_0 is in this state, with zero momentum; they form the condensate. Landau [38] proposed a two-fluid model to explain the phenomena occurring in He II. This model has been remarkably successful in describing the thermal and hydrodynamic properties of helium II.

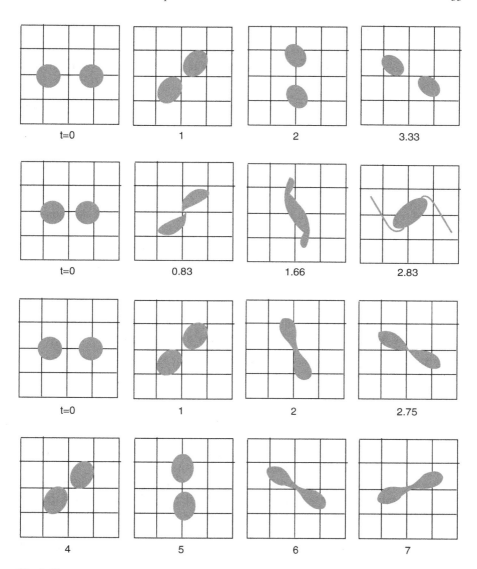

Fig. 1.45

The basic assumptions of the two-fluid model are as follows:

1. He II consists of a kind of mixture of two components – a normal component and a superfluid component. The density of the fluid, ρ, can thus be separated into a normal density ρ_n and a superfluid density ρ_s:

$$\rho = \rho_n + \rho_s. \tag{1.99}$$

Similarly, the fluid motion, characterized by its local velocity $\mathbf{v}$, may be considered to be due to the combined motions of the fluid components, so that

$$\mathbf{J} = \rho\mathbf{v} = \rho_n\mathbf{v}_n + \rho_s\mathbf{v}_s \tag{1.100}$$

where $\mathbf{v}_n$ and $\mathbf{v}_s$ are the velocities of the normal and superfluid components, respectively, and $\mathbf{J}$ is the mass current density.

2. It is assumed that the entropy of the superfluid component is zero ($S_n = 0$), so that the total entropy of the liquid is

$$\rho S = \rho_n S_n$$

3. The superfluid component can move without friction as long as certain velocity limits are not exceeded. The normal component has viscosity.

Quantized vortices: the assertion that superfluid motion is irrotational has been verified experimentally. Both London and Landau arrived at the condition that

$$\nabla \times \mathbf{v}_s = 0.$$

It would follow then that if one rotates a cylindrical vessel full of He II, only the normal component would be carried along by the rotation.

The superfluid component would remain at rest. The height of the meniscus formed in a rotating bucket of He II would be smaller by a factor ρ_n/ρ than the one formed in a classical liquid. But observations by Osborne [76] in 1950 showed that the free superfluid surface assumed the shape of a paraboloid, just as the classical liquid does. On the basis of quantum mechanics Onsager [40] in 1950 showed that there was a possible wave function for the liquid which would produce a motion analogous to classical vortex motion. In fact, he suggested the possibility of quantized vortices in He II. Feynman [77] in 1955 developed this conjecture into a theory which has been successful in explaining many of the strange phenomena observed in the superfluid. His argument was as follows.

Suppose the ground-state wavefunction for the superfluid at rest is ψ_0. Feynman postulated that the wave function, which represents the flowing liquid, has the form

$$\psi_{\text{flow}} = \psi_0 \exp i\Sigma_i W(x_j). \tag{1.101}$$

Then,

$$v_s = \left(\frac{h}{2\pi m}\right)\nabla W, \tag{1.102}$$

where W is a function of position x_j, $W(x_j)$ is its value at the jth atom, and m is the mass of the helium atom. Equation (1.102) implies that

$$\nabla \times v_s = 0.$$

For a simply connected volume of fluid, the circulation around any curve C is then zero:

$$\text{Circulation} = \int_C \left(\frac{dv}{dt}\right) \cdot dl = \int\int_A (\nabla \times v_s) \cdot n dA, \tag{1.103}$$

where A is any surface spanning the curve. For a multiply connected region this is not true, and

$$\int_C v \cdot dl = \int_C \left(\frac{h}{2\pi m}\right)\nabla W \cdot dl = \left(\frac{h}{2\pi m}\right)\int_C dW. \tag{1.104}$$

The circulation is equal to the change in W, the phase of the wave function, in passing around the closed curve C. The superfluid-state wave function must be single-valued, and so

$$\int_C v_s \cdot dl = \frac{nh}{m}, \tag{1.105}$$

where $n = 0, 1, 2, ...$

These singularity lines around which the circulation is not zero are analogous to vortex lines in classical hydrodynamics.

Now, a classical vortex line placed at the origin has a velocity field given by

$$v = \left(\frac{\Gamma}{2\pi r}\right)e_\theta, \tag{1.106}$$

where e_θ is a unit tangential vector, Γ is the vortex strength, and r is the distance from the origin. The streamlines are concentric circles about the vortex line, and the vorticity is zero everywhere except for the vortex line itself.

For a vortex line in He II then,

$$\Gamma = \int_c v \cdot dl = \frac{nh}{m}. \tag{1.107}$$

So, it is possible to have states of motion in the superfluid that are vortex lines, but with the constraint that the circulation about any of these lines must be an integral multiple of h/m. The circulation must obey a quantization condition.

These quantized vortex lines explain how there can be nonzero circulation in a singly connected vessel containing He II while there is zero vorticity throughout nearly all the liquid. This can be demonstrated by analyzing the motion of a superfluid in a rotating bucket. The velocity field per unit length of a vortex parallel to the axis of rotation is

$$v_s = \left(\frac{\Gamma}{2\pi}\right) \times \frac{1}{r}, \quad \Gamma = \frac{hn}{m}. \tag{1.108}$$

The kinetic energy per unit length is

$$E_s = \left(\frac{\rho_s}{2}\right) \int_{r_a}^{r_b} v_s^2 2\pi r \, dr = \rho_s \left(\frac{\Gamma^2}{4\pi}\right) \ln\left(\frac{r_b}{r_a}\right), \tag{1.109}$$

where r_a is the vortex core radius and r_b is a dimension external to the vortex. If the total number of vortices per unit surface area is N, then r_b is given by

$$\pi r_b^2 = \frac{1}{N}. \tag{1.110}$$

Using Stokes' theorem on (1.105),

$$N = \frac{|\mathrm{curl} v_s|}{\Gamma}. \tag{1.111}$$

If F is the free energy of the moving liquid, M the angular momentum, and ω_0 the angular velocity of the vessel, then in the equilibrium state the quantity $F - M\omega_0$ will have a minimum value. The kinetic energy of the motion is

$$\rho_s \left(\frac{v_s^2}{2}\right) + \rho_n \left(\frac{v_n^2}{2}\right).$$

Considering only the motion of the superfluid component, one can minimize the quantity $E_s - M_s\omega_0$. The normal component of the fluid will undergo rotation with a velocity $v_n = \omega_0 r$, just like a solid body. The energy of the superfluid motion is given by

$$E = \left(\frac{1}{2}\right)\rho_s \int v^2 2\pi r dr$$

$$+ \rho_s \left(\frac{\Gamma}{4\pi}\right) \int |\mathrm{curl}\ v| \ln[\Gamma^{1/2}/(\pi^{1/2})|\mathrm{curl} v|^{1/2} r_a)] 2\pi r dr. \tag{1.112}$$

The second term is the energy of the vortices which is obtained by using (1.108–1.111). Similarly, the angular momentum of the fluid is determined to be

$$M = \rho_s \int vr2\pi r dr + \rho_s \left(\frac{\Gamma}{2\pi}\right) \int 2\pi r dr \qquad (1.113)$$

Using a variational technique on the difference $E - M\omega_0$ (see [47], p. 448), the following exact solutions are found:

$$v = \omega_0 r, \qquad (1.114a)$$

$$v = \frac{\text{Constant}}{r}. \qquad (1.114b)$$

Equation (1.114a) implies solid-body rotation and (1.114b) implies irrotational motion with curl $v = 0$.

It is now assumed that solid-body rotation occurs in a region inside the radius R_c. It can be shown [78] that $R - R_c$ is given by

$$R - R_c = \left(\frac{1}{2}\right)\left[\left(\frac{\Gamma}{\omega_0}\right)\frac{\ln r_b}{r_a}\right]^{1/2}. \qquad (1.115)$$

Thus, the region of irrotational motion is small but observable. So, when a bucket of He II is rotated, the vortices formed cause solid-body rotation in nearly the whole bucket. However, in a small region near the wall, there are no vortices and the motion here is completely irrotational.

The same explanation will hold for the shape of the free liquid surface observed.

In 1961, Vinen [79] measured the circulation about a single vortex. He observed that the average circulation of He II around a fine wire along the axis of a slowly rotating cylindrical container was more stable at the value h/m than at any other value. Hess and Fairbank [80] in 1966 showed that the angular momentum of the equilibrium state of a slowly rotating superfluid equals the angular momentum of the lowest free energy state allowed by the quantized vortex model.

It appears then that the quantized vortex concept is correct and accurately describes the states which He II can assume when rotated.

1.4.4.1 Superconductivity

Superconductivity was first identified by Kamerlingh Onnes in 1911 [81], shortly after he had liquefied helium. By the 1960s a satisfactory theoretical concept of classic superconductors had been developed. The subject had to be revisited in 1986 when high-temperature superconductors were discovered by Bednorz and Muller [82]. The primary characteristic of superconductivity is *perfect conductivity*.

The electrical resistance of the material disappears completely within a small temperature range at a critical temperature T_c.

The second is *perfect diamagnetism*, discovered by Meissner and Ochsenfeld in 1933 [83]. They found that a magnetic field is excluded from entering a superconductor. Also, that a field in an originally normal sample is expelled as it is cooled through T_c. The Meissner effect, as it is called, implies that superconductivity will be destroyed by a critical magnetic field H_c.

In 1935, the brothers F. London and H. London [84] proposed two equations to describe the microscopic electric and magnetic fields:

$$\mathbf{E} = \frac{\partial}{\partial t(\Lambda \mathbf{J}_s)}, \tag{1.116}$$

$$\mathbf{h} = -c\,\mathrm{curl}(\Lambda J_s), \tag{1.117}$$

where

$$\Lambda = \frac{4\pi\lambda^2}{c^2} = \frac{m}{n_s e^2}, \tag{1.118}$$

is a phenomenological parameter.

n_s is the number density of superconducting electrons and varies from zero at T_c to a limiting value of the order of n, the density of conducting electrons, at $T \ll T_c$,

$\mathbf{h}$ is the magnetic flux density on a microscopic scale ($\mathbf{B}$ is used for macroscopic values), and $\mathbf{J}_s$ is the flux of electrons.

Equation (1.116) describes perfect conductivity since any electric field *accelerates* the superconducting electrons rather than just maintaining their velocity against resistance (Ohm's law).

Equation (1.117), when combined with the Maxwell equation curl $\mathbf{h} = 4\pi \mathbf{J}/c$, gives

$$\nabla^2 \mathbf{h} = \frac{\mathbf{h}}{\lambda^2}. \tag{1.119}$$

This implies that a magnetic field is exponentially screened from the interior of a material sample with penetration depth λ – the Meissner effect. The parameter is operationally defined as a penetration depth.

Ginzburg–Landau Theory: Ginsberg and Landau proposed a theory of superconductivity as early as 1950 [85] which focused entirely on the superconducting electrons rather than on the excitations. They introduced a complex pseudowavefunction ψ as an order parameter within Landau's general theory of phase transitions. This ψ describes the superconducting electrons, and the local density of superconducting electrons (as defined in the London equations) is

$$n_s = |\psi(x)|^2. \tag{1.120}$$

Using the variational principle, they developed a differential equation for ψ which is analogous to the Schrodinger equation for a free particle, but which has a nonlinear term. The equation for the supercurrent was

$$\mathbf{J}_s = \left(\frac{e*h}{2\pi m*}\right)(\psi * \nabla\psi - \psi\nabla\psi*) - \left(\frac{e*^2}{m*c}\right)|\psi|^2 \mathbf{A}, \tag{1.121}$$

and is the same as the usual quantum mechanical current expression for particles of charge $e*$ and mass $m*$.

With this approach, they were able to treat two features beyond the scope of the London theory:

(a) Nonlinear effects of fields strong enough to change n_s.
(b) The spatial variation of n_s.

The GL theory introduced a characteristic length – the GL coherence length:

$$\xi(T) = \frac{h}{\pi}|2m * \alpha(T)|^{1/2}, \tag{1.122}$$

which gauges the distance over which $\psi(\mathbf{r})$ can vary without significant energy increase.

The ratio of the two characteristic lengths defines the GL parameter, $\kappa = \frac{\lambda}{\xi}$ (1.123)

Since λ also diverges as $(T_c - T)^{-1/2}$ near T_c, this dimensionless ratio is nearly temperature independent.

Type II superconductivity: When λ is greater than ξ, it is energetically favorable for domain walls to form between the superconducting and normal regions. This is called a type II superconductor. When a type II is in a magnetic field, the free energy can be lowered by causing domains of normal material containing trapped flux to form, with low energy boundaries created between the normal core and the surrounding superconducting material. When the applied magnetic field is greater than a value referred to as the lower critical field, B_{c1}, magnetic flux is able to penetrate in quantized units, by forming cylindrically symmetric domains called *vortices*. For applied fields a little above B_{c1}, the magnetic field inside a type II superconductor is strong in the normal cores of the vortices, decreases with distance from the cores, and becomes very small at large distances. For much higher magnetic fields, the vortices overlap and the field inside the superconductor is strong everywhere. When the applied field reaches a value called the upper critical field B_{c2}, the material becomes normal.

1.4.4.2 Vortices

A magnetic field B_{app} will penetrate a superconductor in the mixed state: $B_{c1} < B_{app} < B_{c2}$. Penetration occurs in the form of vortex tubes which confine the flux ([42], Chap. 2). The highest field is in the core which has a radius ξ. The core is surrounded by a region of larger radius λ within which screening currents flowing around the core and magnetic flux are present together.

So, the vortex has a core radius equal to the coherence length ξ and a surrounding outer region with a radius equal to the penetration depth λ. For the high κ limit, $\lambda \gg \xi$ (valid for the copper-oxide superconductors which have typical κ values around 100), the Helmholtz equations can be used (derived from the London formalism) and an expression derived which gives the fraction of the total flux of the vortex that is present in the core [86]:

$$\Phi_{\text{core}} = \frac{\Phi_0}{2\kappa^2} \left(\ln 2\kappa + \frac{1}{2} - \gamma \right). \tag{1.124}$$

Since the magnetic field in the sample is confined to vortices, the total flux is Φ_0 times the number of vortices and the average internal field B_{int} is given by

$$B_{\text{int}} = N_A \Phi_0, \tag{1.125}$$

where N_A is the number of vortices per unit area. For fields much larger than B_{c1} but less than B_{c2}, the internal field is roughly proportional to the applied field and so the density of vortices becomes approximately proportional to the applied field.

The subjects of quantum vortices in superfluids and superconductors will be dealt with in much greater detail later.

1.5 Concluding Remarks

The question of whether vortices play a general role in nature has been considered for many decades. The spiral has stood as a symbol of energy, life, and evolution for centuries. In the biological world, many organs have spiral patterns, for example, the cochlea of the ear [87]. Indeed, the spiral pattern occurs in DNA (in the form of a double helix), the basic molecular building block of all living cells. Krafft [88] has even postulated that life itself is a vortex phenomenon. This may be reaching too far. But what is certain is that the study of the physics of vortices, over the wide size spectrum in which they occur, is very important to our understanding of the physical and biological worlds, and should be recognized as a subject in its own right.

References

1. Homer: The Odyssey. Translated by E.V. Rieu. Penguin Classics (1946)
2. deSantillana, G., von Dechend, H.: Hamlet's Mill. Gambit, Ipswich, MA (1969)
3. Robinson, J.M.: An Introduction to Early Greek Philosophers. Houghton Mifflin, Boston (1968)

4. Fierz, M.: Vorlosungen zur Entwicklungsgeschichte der Mech-anik, Lecture Notes in Physics, vol. 15. Springer, New York (1972)
5. Grant, E. (ed.): A Source Book in Medical Science. Harvard University, Cambridge, MA (1974)
6. Leonardo da Vinci: Del Moto e Mesura deli aqua. In: Carusi, E., Favaro, A. (eds.) Nicola Zanichelli, Bologna (1923)
7. Aiton, E.J.: The Vortex Theory of Planetary Motions. American Elsevier, New York (1972)
8. Whittaker, E.: A History of the Theories of Aether and Electricity, vol. 1. Harper Torchbooks, New York (1960)
9. Lorentz, E.N.: The Nature and Theory of the General Circulation of the Atmosphere. World Meteorological Organization, Geneva (1967)
10. Wegener, A.: Wind-und Wasserhosen. Vieweg, Braunschweig, Germany (1917)
11. Truesdell, C.: Essays in the History of Mechanics. Springer, New York (1968)
12. Schlichting, H.: Boundary Layer Theory. McGraw-Hill, New York (1979)
13. Thompson, W.: Mathematical and Physical Papers, vol. 6. Cambridge University Press, Cambridge (1910)
14. Silliman, R.H.: William Thompson: Smoke Rings and 19th Century Atomism, Isis, vol. 54, p. 461 (1963)
15. Pauli, P.J.: Vortices and Vibrations: The Rise and Fall of a Scientific Research Program, M.A. Thesis, University of Maryland (1975)
16. Serrin, J.: Handbuck der Physik, vol. 8/1, p. 152. Springer, Berlin (1959)
17. Greenspan, H.P.: The Theory of Rotating Fluids. Cambridge University Press, New York (1963)
18. Lin, C.C.: Hydrodynamic Stability. Cambridge University Press, New York (1955)
19. Rosenhead, L.: Laminar Boundary Layers. Oxford University Press, New York (1963)
20. Hinze, J.O.: Turbulence. McGraw-Hill, New York (1975)
21. Goldsten, S.: Fluid mechanics in the first half of this century. Ann. Rev. Fluid Mech. 1, 1–28 (1969)
22. Rotumo, R.: Tornadoes and tornadogenesis. In: Mesoscale Meteorology (1986)
23. Lighthill, M.J.: On sound generated aerodynamically. Proc. Roy. Soc. Lond. A 211, 564 (1952)
24. Powell, A.: Theory of vortex sound. J. Acoust. Soc. Am. 36, 564 (1964)
25. Misner, C.W., Thorne, K.S., Wheeler, J.A.: Gravitation. Freeman, San Francisco, CA (1973)
26. Stone, E.C., Lane, A.L.: Voyager encounter with the jovian system. Science 204, 945 (1979)
27. Stone, E.C., Lane, A.L.: Voyager 2. Science 206, 925 (1979)
28. Problems of Cosmical Aerodynamics. Proceedings of a Symposium on Motion of Gaseous Masses of Cosmical Dimensions. Central Air Documents Office (1951)
29. Slettebak, A. (ed.): Stellar Rotation. Gordon and Breach, New York (1970)
30. Baker, A.: Finite Element CFD. McGraw-Hill, New York (1985)
31. Gunzburger, M., Nicolaides, R. (eds.): Incompressible CFD. Cambridge University Press, New York (1993)
32. Roache, P.: Computational Fluid Dynamics. Hermosa, Albuquerque, New Mexico (1976)
33. Chorin, A.J.: Vortex models and boundary layer instability. SIAM J. Sci. Stat. Comput. 1, 1–21 (1980)
34. Chorin, A.J.: The evolution of a turbulent vortex. Commun. Math. Phys. 35, 517–535 (1982)
35. Leonard, A.: Vortex methods in flow simulation. J. Comput. Phys. 37, 289–335 (1980)
36. Leonard, A.: Computing 3D flows with vortex elements. Am. Rev. Fluid Mech. 17, 523–559 (1985)
37. Tisza, L.: Phys. Rad. 1, p. 165 (1940)
38. Landau, L.: J. Phys. (Moscow) 11, p. 91 (1948)
39. London, F.: Superfluids, vol. 2. Wiley, New York (1954)
40. Onsager, L.: Nuovo Cimento 6, p. 249 (1949)

41. Feynmann, R.P.: In: Porter, C.J. (ed.) Progress in Low Temperature Physics, vol. 1. North-Holland, Amsterdam (1955) (Chapter 2)
42. Belitz, D.: In: Lynn, J.W. (ed.) High Temperature Superconductivity. Springer, Berlin (1990) (Chapter 2)
43. Buzdin, A.J.: Phys. Rev. B **47**, p. 11416 (1993)
44. Lund, F., Regge, T.: Unified approach to strings and vortices with soliton solution. Phys. Rev. D **14**, 1524 (1976)
45. Windmill, S.E.: The structure and dynamics of vortex filaments. Ann. Rev. Fluid Mech. **7**, 141 (1075)
46. Love, A.E.H.: Stabilitat von Ringwirbeln. Encycl. Math. Wiss. Teubner Verlag, Leipzig. Band 4, Teil 3, 86 (1901)
47. McCormack, P.D., Crane, L.: Physical Fluid Dynamics. Academic, New York (1971)
48. von Karman, T., Burgers, J.M.: General aerodynamic theory – perfect fluids. In: Durand, W.F. (ed.) Aerodynamic Theory, vol. 2. Dover, New York (1963)
49. Stananway, S., Shariff, K., Hussan, F.: Head on Collisions of Two Vortex Rings, TR CTR-S88. Center for Turbulence Research, Stanford, CA (1988)
50. Saffman, P.G.: The velocity of a viscous vortex ring. Stud. Appl. Math. **49**, 371–389 (1970)
51. Stananway, S., Cantwell, B.J., Spalart, P.R.: Navier–Stokes Simulations of Axisymmetric Vortex Rings, AIAA Paper No. 88-0318 (1988)
52. Lugt, H.J.: Vortex Flow in Nature and Technology. Wiley, New York (1983)
53. Gibbs-Smith, C.H.: The Invention of the Aeroplane. Taplinger, New York (1966)
54. Wegener, P.P.: What Makes Aeroplanes Fly. Springer, New York (1991)
55. Prandtl, L.: Tragflugeltheorie. Nach. KGes. Wiss. Gottingen, Math. Phys. **Kl**, 15 (1918)
56. Muller, U. et al.: The Wing Section Theory of Kutta and Joukowsky (1979)
57. Breslin, J.A., Adersen, P.: Hydrodynamics of Ships Propellers. Cambridge University Press, New York (1994)
58. Stong, C.L.: Scientific American, October. p. 141 (1971)
59. Shin, H.W., et al.: Circulation measurements and vortical structure in an inlet-vortex flow field. J. Fluid Mech. **162**, 463–487 (1986)
60. Siervi, De, et al.: Mechanisms of inlet-vortex formation. J. Fluid Mech. **174**, 173–207 (1982)
61. Bearman, P.W., Zdravkovich, M.M.: Flow round a circular cylinder near a plane boundary. J. Fluid Mech. **89**, 33–47 (1978)
62. Chen, G.T., et al.: Similarity analysis of compressor tip clearance flow structure. ASME J. Turbomach. **113**, 260–271 (1991)
63. Waitz, I., Greitzer, E., Tan, C.: Fluid vortices. In: Green, S.I. (ed.) Vortices in Aero-Propulsion Systems. Kluwer, Boston, MA (1995). Chapter 11
64. Horlock, J.H.: Axial Flow Compressors. R.E. Krieger, Malabar, FL (1973)
65. von Karman, T., Burgers, J.M.: Motion of a perfect fluid produced by external forces. In: Durand, W.F. (ed.) Sections A2–A5 of General Aerodynamic Theory – Perfect Fluids in Aerodynamic Theory, vol. 2. Dover, New York (1935). Chapter 3
66. O'Connor, N.F.: The Fujihara Effect. Weatherwise. October, p. 232 (1964)
67. Rosenhead, L.: Formation of vortices from a surface of discontinuity. Proc. Roy. Soc. Lond. A **134**, 170 (1931)
68. Roberts, K.V., Christiansen, J.P.: Topics in CFD. In: Macleod, G.R. (ed.) The Impact of Computers on Physics, North-Holland, Amsterdam (1972)
69. Donnelly, R.J.: Experimental Superfluidity. University of Chicago Press (1967)
70. Allen, J.F., Finlayson, D., McCall, D. (eds.): Proceedings of Eleventh International conference on Low Temperature Physics, vol. 1. University of St. Andrews, Scotland (1968)
71. Yarmchuk, E.J., Gordon, M.V., Packard, R.E.: Observations of stationary vortex arrays in rotating superfluid helium. Phys. Rev. Lett. **43**, 214 (1979)
72. Ashton, R.A., Glaberson, W.I.: Phys. Rev. Lett. **42**, p. 1062 (1979)
73. Schwartz, K.W.: Phys. Rev. Lett. **38**, p. 551 (1977)
74. Daunt, J., Mendelssohn, K.: Proc. Roy. Soc. A **170**, p. 423 (1939)

75. Allen, J., Misener, A.: Proc. Roy. Soc. A **173**, p. 467 (1939)
76. Osborne, D.: Proc. Phys. Soc. A **63**, p. 909 (1950)
77. Feynman, R.P.: Atomic theory of the two-fluid model of liquid helium. Phys. Rev. **94**, 262 (1954)
78. Khalatnikov, J.: Introduction to the Theory of Superfluidity. Benjamin and Coy, New York (1965)
79. Vinen, W.: Proc. Roy. Soc. A **260**, p. 218 (1961)
80. Hess, G., Fairbank, W.: Proc. Int. Conf. Low Temp. Phys. 10th, Moscow (1966)
81. Kamerlingh Onnes, H.: Leiden Commun. vol. 120b, 122b, 124c (1911)
82. Bednorz, G., Muller, K.A.: Zeits. Phys. B **64**, p. 189 (1986)
83. Meissner, W., Ochsenfeld, R.: Naturwissenschaften **21**, p. 787 (1933)
84. London, F., London, H.: Proc. Roy. Soc. (Lond.) A **149**, p. 71 (1935)
85. Ginsburg, V.L., Landau, L.: Zh. Eksp. Teor. Fiz. **20**, p. 1064 (1950)
86. Poole, C.P., Farach, H.A., Creswick, R.J.: Superconductivity. Academic, New York (1995). p. 276
87. Cook, T.A.: The Curves of Life. Dover, New York (1979)
88. Kraft, C.F.: Life, a vortex phenomenon. Biodynamics **18**(December), (1936)

Chapter 2
Vorticity (Molecular Spin)

2.1 Introduction

The curl of the fluid velocity vector 7is known as the vorticity or, physically, the angular velocity at a point in space (see [1, p. 68]). It has also been called Rotationgeschwindigkeit by Helmholtz, rotation by Kelvin, molecular rotation by Kelvin, and spin by Clifford.

While vorticity has an exact mathematical definition, its physical significance is still unclear. It is possible, indeed, to find examples of nonrotating flows with nonzero vorticity values. For example, in the shearing motion $u = y$, $v = 0$, and $w = 0$, where u, v, w are the x, y, z components, the particles move in straight lines, but the vorticity or rotation is nonzero.

While circulation is a large-scale measure of rotation, indicative of such features as the Hadley cell in atmospherics, vorticity is a measure of rotation that cannot be seen (microscopic).

Vorticity is the building block of circulation, and the individual locations of vorticity describe pure rotation. However, the sum of vorticity over an area (circulation) is not typically descriptive of pure rotation.

Vorticity, as defined, permits the smooth development of the mathematical study of fluid motion. Stresses within fluids depend on velocity *gradients* rather than on velocities, and vorticity is a combination of velocity gradients. It must be noted that a vorticity field is inherently a solenoidal field: if ζ represents the vorticity vector, then

$$\operatorname{div} \zeta = 0, \tag{2.1}$$

because the divergence of the curl is identically zero. Because the divergence of vorticity is zero, the flux of vorticity out of any closed surface within the fluid is zero. If the closed surface is a vorticity tube, analogous to a stream tube, then the

P. McCormack, *Vortex, Molecular Spin and Nanovorticity: An Introduction*,
SpringerBriefs in Physics, DOI 10.1007/978-1-4614-0257-2_2,
© Percival McCormack 2012

vorticity tubes must either close upon themselves to form a ring or terminate on the boundaries. It appears that the vorticity vector has many of the properties that characterize the velocity vector of an incompressible fluid.

For a particle with zero vorticity, a scalar velocity can be defined such that

$$\mathbf{u} = \nabla \varphi. \tag{2.2}$$

The continuity equation (fluid flow) then gives

$$\nabla \cdot \mathbf{u} = \nabla \cdot (\nabla \phi) = \nabla^2 \phi = 0. \tag{2.3}$$

Thus, ϕ satisfies the Laplace equation.

If the fluid is assumed not to slip on a solid boundary, the relative velocity between the immediately adjacent fluid and the boundary must be zero. If the boundary is stationary, no vorticity tube can terminate on the boundary: so, at the solid stationary wall, there can be no component of vorticity perpendicular to the boundary. Similarly, there can be no component of vorticity parallel to a traction-free boundary.

2.2   Generation of Vorticity

How does a flow that is *initially* irrotational develop vorticity? Truesdell [1] discussed this problem at some length in connection with the Lagrange–Cauchy velocity potential theorem. This theorem deals with the permanence of irrotational motion – that is, a motion once irrotational stays that way. The extension of this theorem to *viscous* incompressible fluids increases the complexity of the problem greatly. Truesdell concluded that "in a motion of a homogenous viscous incompressible fluid subject to a conservative extraneous force and starting from rest, if there be a finite stationary boundary to which the liquid adheres without slipping, there must be some particles whose vorticity is not an analytic function of time at the initial instant."

The measurement of vorticity is difficult but its presence in fluids is easily detected by the determination of circulation, Γ, which is defined as the line integral of the velocity field around any closed curve.

Kelvin's theorem of circulation [2] leads to the following equation:

$$\frac{D\Gamma}{Dt} = \oint -\frac{\nabla p}{\rho} \cdot d\mathbf{l} + \oint \frac{1}{\rho} - \nabla \times (\mu\zeta) + \nabla \left[\left(\frac{4}{3}\right)\mu\nabla \cdot \mathbf{U} \right] \cdot d\mathbf{l} + \oint \mathbf{F} \cdot d\mathbf{l}. \tag{2.4}$$

This shows that the rate of change of circulation within a closed curve l, always made up of the same fluid particles, is governed by the torques produced by pressure forces, body forces, and viscous forces. In (2.4), ρ is the fluid density, p is the fluid pressure, ζ is the vorticity vector, μ is the dynamic viscosity, $\mathbf{U}$ is the velocity vector, and $\mathbf{F}$ is the body force.

Irrotational body forces (conservative body forces with single-valued force potential), present in many fluid dynamic problems, do not produce circulation. Thus, the effects of the body force term can generally be omitted in the treatment of vorticity generation.

Under barotropic conditions ($p = p(\rho$ only)), the density field is a function of the pressure field alone and the pressure forces do not produce circulation. In an inviscid incompressible fluid of uniform density and in an inviscid compressible, provided the flow field is homentropic, internal sources of vorticity do not exist. But, in a compressible, nonhomentropic fluid, pressure forces provide internal sources of vorticity. When all the torque-producing sources are absent, the dynamics of the fluid is governed by Helmholtz's vortex theorems. It is appropriate to recap these theorems at this point.

First theorem: this is general and valid for real flows. It states that the strength of vorticity in a vorticity tube is the same in all cross sections. In addition, a vorticity tube must be closed or must end at a boundary.

Second theorem: valid only for ideal flows of incompressible fluids. It states that vorticity in such a flow can be neither generated nor destroyed, since during the movement of vorticity, the fluid particles cannot leave the vorticity line on which they are positioned.

If the vorticity sources due to pressure are absent, vorticity flux or circulation cannot be created in the interior of the fluid [3]. In such cases, viscosity is generated by viscous forces at a solid boundary or at a free surface.

2.3 Generation by Shock Waves

There are many sources of vorticity in fluid motions and Hadamard [4] was the first to show that vortices are generated by shock waves and that the flow is no longer irrotational after shock waves. This is shown by the Crocco–Vazsonyi equation [5] for the steady flow of an inviscid fluid:

$$\nabla H = T\nabla S + \mathbf{U} \times \zeta, \tag{2.5}$$

where H, T, and S are the total enthalpy, temperature, and entropy, respectively. The vorticity vector, therefore, is dependent on the rates of change of entropy and enthalpy normal to the streamlines. If all streamlines have the same H but different S, there is production of vorticity. This is the situation downstream of a curved shock wave because the entropy increase across a shock wave is determined by the local angle of the shock. Even if the upstream flow is irrotational, the flow downstream of a curved shock wave is rotational.

2.4 Generation by Free Convective Flow and Buoyancy

This is an important source of vorticity in atmospheric and oceanographic flows.

In fact, a free convection flow is produced by buoyancy forces. Temperature differences are introduced, for example, by boundaries maintained at different temperatures and the resulting density differences induce the motion – cold fluids tend to fall, while hot fluids tend to rise. The temperature changes cause variations in the fluid properties – for example, in the density and viscosity. A general analysis is extremely complex, and so some approximation is essential. The Boussinesq approximation is commonly used. In this approximation, variations of all fluid properties other than the density are ignored. Variations in density are considered only insofar as they result in a gravitational force. The continuity equation in its constant density form is

$$\nabla \cdot \mathbf{u} = 0. \tag{2.6}$$

The Navier–Stokes equation is

$$\rho \frac{D\mathbf{u}}{Dt} = -\nabla p + \mu \nabla \mathbf{u} + \mathbf{F}, \tag{2.7}$$

where $\mathbf{F}$ represents the body force term (such forces act on the volume of a fluid particle) and is commonly the effect of gravity, so that $\mathbf{F} = \rho \mathbf{g}$. The gravitational acceleration is derivable from a potential and so $\mathbf{g} = -\nabla \varphi$.

As density variations are important here, $\rho = \rho_0 + \Delta \rho$, and so

$$\mathbf{F} = -(\rho_0 + \Delta \rho)\nabla \phi = -\nabla(\rho_0 \phi) + \Delta \rho \mathbf{g}. \tag{2.8}$$

Introducing, $P = p + \rho_0 \phi$, then the Navier–Stokes equation becomes

$$\rho \frac{D\mathbf{u}}{Dt} = -\nabla P + \mu \nabla^2 \mathbf{u} + \Delta \rho \mathbf{g}. \tag{2.9}$$

If it is assumed that all accelerations in the flow are small compared to $|\mathbf{g}|$, then the dependence of ρ on T can be linearized:

$$\Delta \rho = -\alpha \rho_0 \Delta T, \tag{2.10}$$

where α is the coefficient of expansion of the fluid. The Boussinesq dynamical equation then is

$$\frac{D\mathbf{u}}{Dt} = \left(\frac{-1}{\rho}\right)\nabla p + \nu \nabla^2 \mathbf{u} - \mathbf{g}\alpha \Delta T, \tag{2.11}$$

where the normal ρ and p have been reverted to.

An equation for the temperature is also required.

To be consistent with the Boussinesq approximation, it is postulated that the fluid has a constant heat capacity per unit volume, ρC_P. $\rho C_P DT/Dt$ is the rate of heating per unit volume of a fluid particle. The heating is caused by transfer of heat from nearby fluid particles by thermal conduction and can also be due to internal heat generation. The corresponding terms in the thermal equation are analogous to the viscous term and the body force term in the dynamical equation, respectively. The conductive heat flux is

$$\mathbf{H} = -k\,\mathrm{grad}\,T, \tag{2.12}$$

where k is the thermal conductivity of the fluid. Thus,

$$\rho C_P \frac{DT}{Dt} = -\mathrm{div}\,\mathbf{H} + J, \tag{2.13}$$

where J is the rate of internal heat generation per unit volume. Taking k to be a constant, (2.13) can be modified to

$$\frac{\partial T}{\partial t} + \mathbf{u} \cdot \nabla T = \kappa \nabla^2 T + \frac{J}{\rho C_P}, \tag{2.14}$$

where $\kappa = k/\rho C_P$ is the thermal diffusivity.

Equations (2.6), (2.11), and (2.14) are the basic equations of convection in the Boussinesq approximation.

The additional term in the dynamical equation, $-\mathbf{g}\alpha\Delta T$, is known as the buoyancy force. The two terms on the right-hand side of (2.14) are the conduction term and heat generation term, respectively. $\mathbf{u}\cdot\nabla T$ is known as the advection term (transport of heat by the motion).

Equation (2.14) requires boundary conditions for the temperature field.

The most common type specifies the wall temperature, which is the temperature of the fluid in contact with the wall.

It must be noted that thermal conduction plays an integral role in convection.

A wide range of fluid dynamical behavior is to be expected, depending on the importance of the buoyancy force with respect to the other terms in (2.11). When the buoyancy force is negligible, one has forced convection – when it is the only cause of motion, there is free convection.

Free convective flows are normally rotational. Buoyancy forces directly generate vorticity. Applying the curl operation to (2.11):

$$\frac{D\zeta}{Dt} = \zeta \cdot \nabla \mathbf{u} + \nu \nabla^2 \zeta + \alpha \mathbf{g} \times \nabla(\Delta T). \tag{2.15}$$

This is just the vorticity equation, which will be dealt with in the next section, with the addition of the buoyancy force.

Horizontal components of the temperature gradient $\nabla(\Delta T)$ contribute to the last term; the vorticity so generated is also horizontal but perpendicular to the temperature gradient.

2.5   Generation by Baroclinic Effects

Kelvin's theorem for a potential body force and inviscid flow has the simple form:

$$\frac{D\Gamma}{Dt} = -\oint \frac{dp}{\rho}. \tag{2.16}$$

If the flow is baroclinic (with lines of constant ρ not parallel to lines of constant p), the baroclinicity will be a source of vorticity.

Note: in general, a barotropic situation is one in which surfaces of constant pressure and surfaces of constant density coincide; a baroclinic situation is one in which they intersect.

The sea-breeze problem [6] illustrates the baroclinic generation of circulation. A temperature difference between the air over the land and that over the sea generates density differences between the land air mass and the sea air mass. Thus, the flow isobars and the flow isopycnals are not coincident. For the typical situation shown in Fig. 2.1, Green [7] has shown that for a 15°C land–sea temperature difference ($\rho_2 = 1.22$ kg/m^3 and $\rho_1 = 1.18$ kg/m^3) and for an elevation change of 1 km ($p_0 - p_1 = 12$ kPa), $D\Gamma/Dt = (p_0 - p_1)(1/\rho_2 - 1/\rho_1) = -333$ m^2/s. The mean tangential velocity is evaluated from $\Gamma = v_{\text{mean}} \times (h + L)$ and it is found that

$$\frac{dv_{\text{mean}}}{dt} \approx \frac{-333 \, (\text{m}^2/\text{s}^2)}{62,000 \, (\text{m})} = 20 \, \text{m/s/h}.$$

Thus, this simple model predicts that in 1 h baroclinicity will induce a sea breeze of 20 m/s (40 knots) from the sea to land at low elevation and vice versa at high elevation. In atmospheric flows, then, baroclinicity can be an important generator of vorticity.

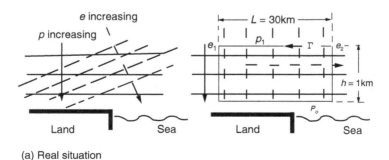

(a) Real situation

Fig. 2.1 Flow dynamics due to temperature differences between land and sea air masses

2.6   The General Vorticity Equation

The first step is to use the momentum equation written in an inertial frame of reference:

$$\frac{D\mathbf{u}}{Dt} = \frac{\partial \mathbf{u}}{\partial t} + (\mathbf{u} \cdot \nabla)\mathbf{u} = \frac{-\nabla p}{\rho} + \frac{\nabla \cdot \mathbf{T}}{\rho} + \mathbf{g}, \tag{2.17}$$

where $\mathbf{T}$ is the stress tensor and $\mathbf{g}$ is the body force per unit mass.

Now, $(\mathbf{u} \cdot \nabla)\mathbf{u} = \nabla((1/2)\mathbf{u} \cdot \mathbf{u}) - \mathbf{u} \times \boldsymbol{\omega}$ and so (2.17) becomes

$$\frac{\partial \mathbf{u}}{\partial t} + \nabla\left(\frac{\mathbf{u} \cdot \mathbf{u}}{2}\right) - \mathbf{u} \times \omega = \frac{-\nabla p}{\rho} + \frac{\nabla \cdot \mathbf{T}}{\rho} + \mathbf{g}. \tag{2.18}$$

The curl of this equation, after some simplification, is

$$\frac{D\omega}{Dt} = -\omega(\nabla \cdot \mathbf{u}) + (\omega \cdot \nabla)\mathbf{u} + \left(\frac{1}{\rho^2}\right)\nabla \times \nabla p - \left(\frac{1}{\rho^2}\right)\nabla\rho \times (\nabla \cdot \mathbf{T})$$

$$+ \left(\frac{1}{\rho}\right)\nabla \times (\nabla \cdot \mathbf{T}) + \nabla \times \mathbf{g}. \tag{2.19}$$

Using the momentum equation to eliminate ∇p, another form of the general form of the vorticity equation is

$$\frac{D\omega}{Dt} = -\omega(\nabla \cdot \mathbf{u}) + (\omega \cdot \nabla)\mathbf{u} + \left(\frac{1}{\rho}\right)\nabla\rho \times \left(\mathbf{g} - \frac{D\mathbf{u}}{Dt}\right) + \left(\frac{1}{\rho}\right)\nabla \times (\nabla \cdot \mathbf{T}) + \nabla \times \mathbf{g}.$$

$$\tag{2.20}$$

It must be noted here that the angular velocity vector $\boldsymbol{\omega}$ is being used synonymously with the vorticity vector symbol ζ used above. This can be confusing at times.

The physical significance of the terms in (2.19) is very important. The left-hand side is the time rate of change of the vorticity following a specific fluid element – the convective transport of vorticity.

The first term on the right-hand side represents the reduction in vorticity due to fluid expansion. Vorticity is enhanced by fluid compression and this will be discussed in some more detail later. The next term represents a stretching of vortex lines that intensifies the vorticity. By Kelvin's theorem, the total circulation of the vortex lines must be constant and so the axial stretching of vorticity lines increases their vorticity. In terms of angular momentum, it is known that a thick solid rod spinning about its axes on frictionless bearings spins faster when stretched in order to preserve angular momentum – see Fig. 2.2. Tornados are highly stretched vortices resulting in powerful winds and are a good example of such vorticity intensification.

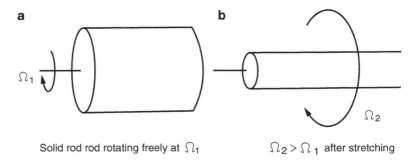

Solid rod rod rotating freely at Ω_1 $\Omega_2 > \Omega_1$ after stretching

Fig. 2.2

The third term on the right-hand side is the "baroclinic torque" caused by the noncollinearity of the density and pressure gradients. This will be discussed further below. The fourth term is due to shear stress variations in a density gradient field resulting in a torque. In engineering flows it is neglected, being much smaller than the other terms; but it cannot be neglected in meteorological flows.

The fifth term represents the diffusion of vorticity due to viscosity and also will be treated in more detail later.

In engineering, one typically deals with body forces, such as gravity, which are potential, and so the last term on the right-hand side would be zero.

As defined previously, the vorticity, $\boldsymbol{\omega}$, of a flow field with velocity distribution $\mathbf{u}$ is

$$\omega = \nabla \times \mathbf{u}. \tag{2.21}$$

Equations (2.19) and (2.20) and the continuity equation

$$\frac{D\rho}{Dt} + \rho \nabla \cdot \mathbf{u} = 0, \tag{2.22}$$

should be sufficient to determine the vorticity field everywhere. But (2.20) does not contain the pressure field. The equation necessary to find p is generated by taking the divergence of (2.18) above:

$$\frac{\partial (\nabla \cdot \mathbf{u})}{\partial t} + 2(\mathbf{u} \cdot \nabla)(\nabla \cdot \mathbf{u}) + \nabla^2 \left(\frac{\mathbf{u} \cdot \mathbf{u}}{2}\right) - \mathbf{u} \cdot (\nabla^2 \mathbf{u}) - \omega \cdot \omega$$

$$= \left(\frac{-1}{\rho}\right) \nabla^2 p + \left(\frac{1}{\rho}\right) \nabla \cdot (\nabla \cdot \mathbf{T}) + (\nabla \cdot \mathbf{T} - \nabla p) \cdot \nabla \left(\frac{1}{\rho}\right) + \nabla \cdot \mathbf{g}. \tag{2.23}$$

In the case of a constant density fluid ($\nabla \cdot \mathbf{u} = 0$ by continuity) and constant viscosity (then $\nabla \cdot \mathbf{T} = \mu \nabla^2 \mathbf{u}$) this very complex equation reduces to

$$\left(\frac{1}{\rho}\right) \nabla^2 p = \mathbf{u} \cdot (\nabla^2 \mathbf{u}) + \omega \cdot \omega - \nabla^2 \left(\frac{\mathbf{u} \cdot \mathbf{u}}{2}\right) + \nabla \cdot \mathbf{g}. \qquad (2.24)$$

Once the vorticity and velocity fields are determined using (2.20), the pressure field is given by (2.24).

2.7 Viscous Diffusion of Vorticity

Here, the flow of a Newtonian fluid of fixed density and viscosity with only potential body forces will be considered.

Equation (2.19) reduces to

$$\frac{D\omega}{Dt} = (\omega \cdot \nabla)\mathbf{u} + \nu \nabla^2 \omega. \qquad (2.25)$$

This has the form of a convection–diffusion equation, similar to the equations of thermal convection and substance diffusion. It explicitly implies the diffusion of vorticity due to the action of viscosity.

The "Oseen" or "Lamb" vortex is a classic example of this diffusion phenomenon. Oseen [8] has analyzed the case of a two-dimensional axisymmetric line vortex in an initially inviscid, infinite fluid. From time $t = 0$, the viscosity acts and the resulting diffusion of vorticity is analyzed. To facilitate interpretation, (2.25) is expanded as

$$\frac{\partial \omega}{\partial t} + (\mathbf{u} \cdot \nabla)\omega = (\omega \cdot \nabla)\mathbf{u} + \nu \nabla^2 \omega. \qquad (2.26)$$

Following Truesdell's terminology, the terms in this equation are given specific physical identities. The term $\partial \omega / \partial t$ is called the diffusion of vorticity. $\mathbf{u} \cdot \text{grad}\omega$ is called the convection of vorticity. $\omega \cdot \text{grad}\mathbf{u}$ is the convective rate of change of vorticity, and $D\omega/Dt$ is the diffusive rate of change of vorticity. Finally, the term $\nu \nabla^2 \omega$ is called the dissipative rate of change of vorticity.

The Oseen vortex is two dimensional and so the third term is zero (ω is normal to the plane of $\nabla \mathbf{u}$). The second term is zero for similar reasons and so

$$\frac{\partial \omega_z(r, t)}{\partial t} = \nu \nabla^2 \omega_z(r, t).$$

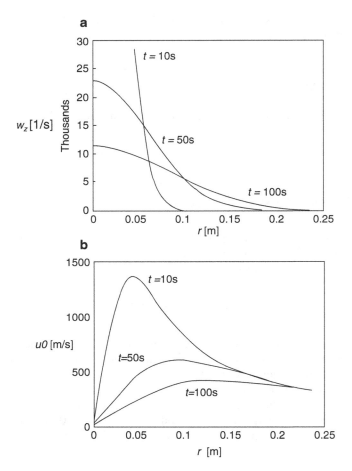

Fig. 2.3 (a) Vorticity distribution. (b) Resulting tangential velocity

This Poisson equation is readily solved and the solution is

$$\omega_z(r,t) = \left(\frac{\Gamma_0}{4\pi vt}\right)\exp\left(\frac{-r^2}{4vt}\right). \tag{2.27}$$

Figure 2.3a [7] shows the vorticity distribution for a circulation $\Gamma_0 = 500\,\mathrm{m^2/s}$ and kinematic viscosity $v = 3.5 \times 10^{-5}\,\mathrm{m^2/s^2}$. These values are typical for the tip vortex generated by a large aircraft at cruising altitudes.

Figure 2.3b shows the resulting tangential velocity, $u_\theta = (\Gamma_0/2\pi r)[1 - \exp(-r^2/4vt)]$. There are three features of Fig. 2.3 which should be specially noted:

1. The circulation distant from the centerline is time independent as per Kelvin's theorem.

2. Distant from the vortex centerline the vorticity and velocity distributions are unchanged. It takes considerable time for viscosity to alter the vorticity over long distances.
3. Near the center of the vortex, where the velocity gradient is largest, the motion becomes a solid-body rotation quickly. Viscosity always acts to bring to convert vortex cores into solid-body rotation.

2.7.1 Viscous Diffusion at a Wall

Diffusion of vorticity is an exchange of momentum. Transport of vorticity can also occur by convection with the property that vorticity is preserved on a particle path. Thus, vorticity can be transferred to neighboring paths only by diffusion – that is, by the effect of viscosity. Immediately at a wall, to which the fluid particles adhere, vorticity can be transferred to the fluid only by diffusion. Boundary layers at surfaces have large velocity gradients and result in large viscous forces. Vorticity generation and diffusion at walls will be next considered.

Assuming constant density and constant Newtonian viscosity, and with some manipulation, (2.17) may be written as

$$\frac{D\mathbf{u}}{Dt} = \frac{-\nabla p}{\rho} + \mathbf{g} - \nu \nabla \times \omega. \tag{2.28}$$

For two-dimensional flow $(u = u(x,y), v = v(x,y), w = 0)$ near a wall $(\boldsymbol{\omega} = \omega_z$ only and $u(\text{wall}, y = 0) = 0)$, and neglecting body forces, then the x-component of the momentum equation at the wall is

$$\frac{\nu \partial u}{\partial y} + \left(\frac{1}{\rho}\right)\frac{\partial p}{\partial z} = \frac{-\nu \partial \omega_z}{\partial y}. \tag{2.29}$$

There are two main implications of (2.29):

1. Lighthill [9] has identified the term $\nu(\partial \omega_z/\partial y)$ as the flux of vorticity away from a surface, with positive fluxes representing the flux of positive vorticity and negative fluxes representing the flux of vorticity of opposite sign [10]. Thus, only solid surfaces with pressure gradients, and porous walls with fluid blowing out, are sources of vorticity. In the Blasius boundary layer (formed by fluid flowing over a thin flat plate), in fact, all the vorticity in the boundary layer is generated in the small leading edge region where $\partial p/\partial x$ is negative (see Fig. 2.4). Vorticity in other regions of the Blasius boundary layer is due to convection from the leading edge.
2. Without a pressure gradient at the wall $(\partial p/\partial x)$, or flow through it $(v = 0)$, then $(\partial \omega_z/\partial y)_{y=0} = 0$, which implies that vorticity generated at the wall has diffused far into the flow(Fig. 2.4a). A porous wall with suction has $v(y = 0) < 0$

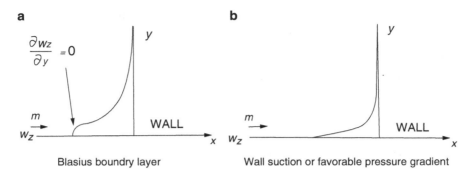

Fig. 2.4

(Fig. 2.4b). Equation (2.29) then implies $\partial\omega_z/\partial y_{y=0} > 0$. Thus, diffusion of wall vorticity into the flow has been inhibited by the wall suction. Contrariwise, blowing at a wall, or an adverse pressure gradient along a wall, will result in vorticity transport away from the wall. Batchelor [3] has solved the problem of suction at a wall quantitatively. Steady incompressible two-dimensional flow over a flat plate with suction velocity V, with no pressure gradients or body forces, must satisfy the following vorticity equation:

$$\frac{-V\mathrm{d}\omega_z}{\mathrm{d}y} = \frac{v\mathrm{d}^2\omega_z}{\mathrm{d}y^2}. \tag{2.30}$$

The solution that satisfies the no slip ($u(y = 0) = 0$) and freestream ($u(y \to \infty)$ $= U_\infty$) conditions is

$$\omega_z = \left(\frac{-U_\infty V}{v}\right)\varepsilon^{-Vy/v} \quad \text{and} \quad u = U_\infty(1 - \varepsilon^{-Vy/v}). \tag{2.31}$$

This indicates that as the suction velocity increases, the thickness of the region (of the order v/V) with vorticity decreases. In the example described above, the convection of vorticity through the wall exactly compensates for diffusion into the free stream, resulting in a streamwise invariant flow. If $V = 0$, then a streamwise invariant boundary-layer flow would only be possible with a favorable pressure gradient.

A favorable pressure gradient will also inhibit vorticity diffusion into the free-stream. Boundary-layer flow near a stagnation point has a highly favorable pressure gradient and this problem has been solved numerically [11]. It was determined

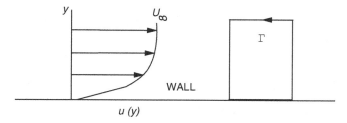

Fig. 2.5 Wall-generated vorticity

that the boundary-layer thickness, δ, in which $\boldsymbol{\omega}$ is significant, is proportional to $(\nu/U_\infty)^{1/2}$, where U_∞ is the far field fluid velocity toward the stagnation point. The "vorticity layer" thickness does not vary with distance away from the stagnation point along the plate. Thus, the large favorable pressure gradient in effect halts the diffusion of vorticity from the wall, in spite of producing a surface vorticity flux.

2.7.2   Subsequent Motion of Wall-Generated Vorticity

Kelvin's theorem implies that the flow region away from viscous effects remains irrotational for all time and may be computed by potential flow methods. Thus, determination of the complete flowfield around an object will require knowing the amount and motion of the wall-generated vorticity. As an approximate model of a boundary layer, consider a one-dimensional flow over a surface – see Fig. 2.5. Assuming that convection is the only mechanism for vorticity (valid at reasonably high Reynolds numbers because convection is a much faster process than diffusion), then the amount of vorticity convected past a fixed vertical line in time $\mathrm{d}t$ is

$$\Gamma = \int_0^\infty \mathrm{d}t \cdot u(y)\omega_z(y)\mathrm{d}y. \tag{2.32}$$

For one-dimensional flow, $\omega_z(y) = -\mathrm{d}u/\mathrm{d}y$ and, therefore,

$$\Gamma = -\mathrm{d}t \int_0^\infty u\left(\frac{\mathrm{d}u}{\mathrm{d}y}\right)\mathrm{d}y = -\mathrm{d}t \int_{u(0)}^{u(\infty)} u\,\mathrm{d}u = -\mathrm{d}t \left(\frac{u^2}{2}\right)\Big|_0^{U_\infty} = -U_\infty\left(\frac{U_\infty}{2}\right)\mathrm{d}t. \tag{2.33}$$

This result is important as it indicates how much circulation must be injected into the flow at boundary-layer separation points. This is required for vortex method computations.

Fig. 2.6

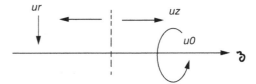

2.7.3 Vorticity Increase by Vortex Stretching

This has already been referred to as another consequence of Kelvin's theorem. Burgers' vortex is an example of this phenomenon and will be analyzed next. For an inviscid barotropic flow with only potential body forces, (2.19) reduces to

$$\frac{D\omega}{Dt} = \omega \cdot \nabla \mathbf{u} - \omega(\nabla \cdot \mathbf{u}). \tag{2.34}$$

Using the continuity equation, $\nabla \cdot \mathbf{u} = (-1/\rho)(D\rho/Dt)$ and the product rule

$$D\left(\frac{\omega}{\rho}\right) = \left(\frac{1}{\rho}\right)\left(\frac{D\omega}{Dt}\right) - \left(\frac{\omega}{\rho^2}\right)\left[\frac{D\rho}{Dt}\right] : \frac{D(\omega/\rho)}{Dt} = \left(\frac{\omega}{\rho}\right) \cdot \nabla \mathbf{u}. \tag{2.35}$$

Now, the length of an infinitesimal segment, $\mathbf{l}$, of a fluid line [7, p. 12] is given by

$$\frac{D\mathbf{l}}{Dt} = \mathbf{l} \cdot \nabla \mathbf{u}. \tag{2.36}$$

From (2.35) and (2.36) it is seen that for an arbitrary flow field

$$\mathbf{l} = \frac{C\omega}{\rho}, \tag{2.37}$$

where C is a constant. See [3] for a more rigorous derivation.

Equation (2.37) shows that by stretching a segment of fluid with vorticity, so that $\mathbf{l}$ increases, the vorticity magnitude of the segment will also increase. It is also obvious that fluid compression (increase in ρ) will also lead to vorticity augmentation.

Burgers [12] has solved a problem (the Burger vortex) that clearly shows the increase in vorticity by stretching. The vorticity equation in an incompressible, Newtonian zero body force fluid is [see (2.26)]

$$\frac{D\omega}{Dt} = (\omega \cdot \nabla)\mathbf{u} + v\nabla^2\omega. \tag{2.38}$$

For an axisymmetric vortex, aligned along the z-axis and placed in a uniaxial straining field along its length (see Fig. 2.6) $u_z = 2Cz$ (where C is a constant), then continuity demands the presence of a radial influx of fluid $u_r = -Cr$.

Equation (2.38) has a Lamb–Oseen vortex type of solution:

$$u_z = 2C_z$$

$$u_r = -Cr$$

$$u_\theta = \left(\frac{\Gamma}{2\pi r}\right)\left[1 - \exp\left(\frac{-r^2}{4\delta^2}\right)\right], \qquad (2.39)$$

where

$$\delta^2 = \left(\frac{v}{C}\right) + \left(\delta_0^2 - \frac{v}{C}\right)\exp(-Ct). \qquad (2.40)$$

δ may be defined as the vortex radius and δ_0 is then the initial vortex radius. This velocity field has a vorticity distribution given by

$$\omega_z = \left(\frac{\Gamma}{\pi\delta^2}\right)\exp\left(\frac{-r^2}{4\delta^2}\right). \qquad (2.41)$$

The vorticity on the axis is given by $\omega_z(r = 0) = \Gamma/\pi\delta^2$.

From (2.40) $\delta^2 \rightarrow v/C$ as t increases (for C positive). If the vortex is being compressed, C negative, δ increases continuously. If δ_0 is greater than $(v/C)^{1/2}$, the asymptotic value of δ, δ will decrease with time. For this situation $\omega_z(r = 0)$ will rapidly increase with time. This is a good example of the increase in vorticity that results from vortex stretching.

2.8   Hill's Spherical Vortex [13, 14]

This is another good example in which stretching of vortex lines occurs. The spherical vortex is a model of the internal flow in a gas bubble moving in a liquid. The motion outside the bubble sets up an internal circulation. Figure 2.7 shows the spherical vortex, with a cylindrical coordinate system moving with the bubble.

The vortex lines are circular loops around the z-axis. As the flow moves the vortex lines to larger radial positions, the loops increase in length proportional to the radius. As a result of the vortex-line stretching effect, the vorticity is proportional to the radius (see [14] for the value of C as $5U/R^2$):

$$\omega_\theta = Cr = \left(\frac{5U}{R}\right)\frac{r}{R}. \qquad (2.42)$$

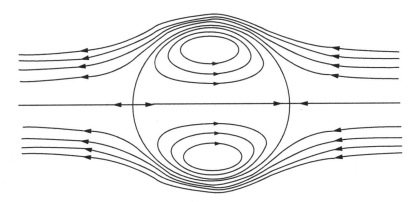

Fig. 2.7 Hill's spherical vortex

Consider the vorticity equation (2.38) as it applies to Hill's vortex and recall that $D\boldsymbol{\omega}/Dt$ is the rate of change of particle vorticity, $\boldsymbol{\omega}\cdot\nabla\mathbf{u}$ is the rate of deforming of vortex lines, and $\nu\nabla^2\boldsymbol{\omega}$ is the net rate of viscous diffusion of $\boldsymbol{\omega}$.

Only the θ component has nonzero vorticity. The physical meaning of the terms in (2.38) is then

Convection: $D\omega_\theta/Dt = u_r \partial\omega_\theta/\partial r = Cu_r$
Stretching: $[\omega \cdot \nabla\mathbf{u}] = \omega_\theta(u_r/r) = Cu_r$
Diffusion: $[\nu\nabla^2\omega]_\theta = \nu\partial/\partial r[(1/r)\partial/\partial r(r\omega_\theta)] = 0$

The vorticity balance then is between convection and stretching without any net viscous diffusion. There is no net diffusion of vorticity and the increase in vorticity is wholly due to vortex-line stretching. The vorticity is proportional to the circumference of the loop and does not depend on the movement of the circular vortex lines.

2.9 Vorticity in Rotating Frames of Reference

Kelvin's theorem for a potential force and inviscid flow is given in (2.16). This equation and other results developed so far are valid only in an inertial frame of reference. In studying atmospheric and oceanic flows, one deals with the noninertial, rotating frame of the Earth. It is appropriate to reformulate (2.16) in such a rotating frame.

Two additional forces occur in a rotating frame of reference (with rotational velocity $\boldsymbol{\Omega}$) – the centrifugal force and the Coriolis force.

The centrifugal force, $\mathbf{F}_{\text{cen}} = \boldsymbol{\Omega} \times (\boldsymbol{\Omega} \times \mathbf{x})$ per unit mass, where $\mathbf{x}$ is the vector displacement from the axis of rotation, generates no circulation because $\mathbf{F}_{\text{cen}} =$

$\nabla(1/2\mathbf{\Omega} \times \mathbf{x})^2$ is curl free. Coriolis forces, $\mathbf{F}_{cor} = -2\mathbf{\Omega} \times \mathbf{u}$ per unit mass, are not curl free and do generate circulation. This circulation can be calculated as follows.

An area of fluid A normal to the axis of rotation, in the rotating frame, has a circulation (about an axis parallel to the rotation) in the stationary frame of reference given by the rigid body rotation: $\Gamma = 2|\mathbf{\Omega}|A$. If the area A has a normal at an inclination $\pi/2 - \phi$ to the axis of rotation (at the North Pole on Earth $\phi = \pi/2$ and at the South Pole $\phi = -\pi/2$), the area that determines the rotating frame circulation is the projected area of A on the equatorial plane, $A_e = A \sin\phi$.

Thus, the circulation in a rotating frame of reference is given by

$$\Gamma_{rot} = 2|\mathbf{\Omega}|A\sin\phi. \tag{2.43}$$

This circulation is added to the circulation calculated relative to the rotating frame of reference, Γ_{rel}, to yield the circulation in the absolute frame of reference:

$$\Gamma_{ab} = \Gamma_{rot} + \Gamma_{rel} = 2|\mathbf{\Omega}|A\sin\phi + \Gamma_{rel}. \tag{2.44}$$

Substituting (2.44) into (2.16), one obtains Bjerknes [15] theorem:

$$\frac{D\Gamma_{rel}}{Dt} = -\int \frac{dp}{\rho} - 2|\mathbf{\Omega}|\frac{d(A\sin\phi)}{dt}. \tag{2.45}$$

Besides the baroclinic torque term discussed previously, there is a new source of vorticity convection of fluid from high latitudes to low latitudes. Consider a part of the atmosphere on the Earth at a latitude ϕ_1, about which there is no relative circulation. Suppose the air is now brought barotropically to latitude ϕ_2 with no change in area, then the mean vorticity of the air mass will alter as

$$\frac{AD\omega}{Dt} = \frac{-2|\mathbf{\Omega}|Ad\sin\phi}{dt} \quad \text{or} \quad \omega_2 - \omega_1 = -2|\mathbf{\Omega}|(\sin\phi_2 - \sin\phi_1). \tag{2.46}$$

Counterclockwise fluid rotation ($+\omega$) is therefore enhanced in the Northern Hemisphere as fluid moves south and the reverse occurs in the Southern Hemisphere. Winds generated by this motion of the air equatorward across lines of latitude are called "cyclonic" winds, and are associated with low-pressure regions in the atmosphere. Coriolis force-generated circulation also plays an important part in oceanographic currents and in turbomachinery.

2.10 Atmospheric Fluid Motion and Vorticity

Analogous to absolute circulation, there is the absolute vorticity – the sum of the vorticity due to the rotation of the fluid itself (ζ) and that due to the Earth's rotation (f). f is known as the Coriolis parameter and varies only with latitude. Under the conditions of nondivergent, frictionless flow, absolute vorticity is conserved and

$$\frac{\mathrm{d}(\zeta + f)}{\mathrm{d}t} = 0. \tag{2.47}$$

For two-dimensional flow, in an atmosphere of uniform density, and conservation of absolute vorticity, the vorticity equation becomes

$$\frac{\mathrm{d}}{\mathrm{d}t}(\zeta + f) = -(\zeta + f)\left(\frac{\partial u}{\partial x} + \frac{\partial v}{\partial y}\right) - \left(\left(\frac{\partial w}{\partial x}\right)\left(\frac{\partial v}{\partial z}\right) - \left(\frac{\partial w}{\partial y}\right)\left(\frac{\partial u}{\partial z}\right)\right)$$

$$+ \left(\frac{1}{\rho^2}\right)\left(\left(\frac{\partial r}{\partial x}\right)\left(\frac{\partial p}{\partial y}\right) - \left(\frac{\partial r}{\partial y}\right)\left(\frac{\partial p}{\partial x}\right)\right). \tag{2.48}$$

The first term on the right-hand side arises because of the horizontal divergence. If there is a positive horizontal divergence, air flows out of the region in question and the vorticity will decrease. This is equivalent to the situation with a rotating body whose angular velocity decreases when its moment of inertia increases (angular momentum conservation). For synoptic scale (systems of 1,000 km or more in horizontal dimension) motions, the last two terms are much smaller than the others and to a first approximation

$$\frac{\mathrm{d}_h}{\mathrm{d}t}(\zeta + f) = -(\zeta + f)\left(\frac{\partial u}{\partial x} + \frac{\partial v}{\partial y}\right), \tag{2.49}$$

where $\mathrm{d}_h/\mathrm{d}t$ denotes $\partial/\partial t + u\partial/\partial x + v\partial/\partial y$.

Applying (2.49) to a constant density and temperature atmosphere using the continuity equation for incompressible fluids, it becomes

$$\frac{\mathrm{d}_h}{\mathrm{d}t}(\zeta + f) = (\zeta + f)\frac{\partial w}{\partial z}. \tag{2.50}$$

Because of the constant temperature, the geostrophic (small friction, small curvature, and steady flow) wind is independent of the height z. The vorticity will not vary with height either, because to a first approximation the vorticity is equal to the vorticity of the geostrophic wind. Then, integrating (2.50) between levels z_1 and z_2 where $z_2 - z_1 = h$,

$$\frac{(1/(\zeta + f))\mathrm{d}_h}{\mathrm{d}t(\zeta + f)} = \frac{w(z_2) - w(z_1)}{h}. \tag{2.51}$$

Considering the fluid which at one instant is confined between the levels distance h apart, then $\mathrm{d}h/\mathrm{d}t = w(z_2) - w(z_1)$ and (2.50) may be written as

$$\frac{\mathrm{d}_h}{\mathrm{d}t(\zeta + f/h)} = 0. \tag{2.52}$$

This equation is a simplified statement of the conservation of *potential vorticity*. It has important consequences for atmospheric flow. For example, consider adiabatic flow over a mountain barrier. As a column of air flows over the mountain, its vertical extent is decreased and so ζ must also decrease. A westward-moving wind will therefore move in the direction of the equator as it flows over the mountain.

2.11 Dissipation Function, Vorticity Function, and Curvature Function (Eddy or Vortex Motion)

Following Lamb [16], the rate of dissipation is defined as the energy expended in deforming a small fluid element and is mathematically defined by the dissipation function, Φ. For an incompressible fluid, in rectangular Cartesian coordinates,

$$\Phi = \mu \left[2\left(\frac{\partial u}{\partial x}\right)^2 + 2\left(\frac{\partial v}{\partial y}\right)^2 + 2\left(\frac{\partial w}{\partial z}\right)^2 + \left(\frac{\partial w}{\partial y} + \frac{\partial v}{\partial z}\right)^2 + \left(\frac{\partial u}{\partial z} + \frac{\partial w}{\partial x}\right)^2 + \left(\frac{\partial v}{\partial x} + \frac{\partial u}{\partial y}\right)^2 \right].$$

$$(2.53)$$

A vorticity function, Ω, may be defined by taking the scalar product of the vorticity vector with itself. This function is used as a measure of vorticity:

$$\Omega = \omega \cdot \omega = \left(\frac{\partial w}{\partial y} - \frac{\partial v}{\partial z}\right)^2 + \left(\frac{\partial u}{\partial w} - \frac{\partial w}{\partial x}\right)^2 + \left(\frac{\partial v}{\partial x} - \frac{\partial u}{\partial y}\right)^2, \qquad (2.54)$$

where ω is the vorticity vector. It can be shown [17] that Φ and Ω are mathematically independent functions; that is, dissipation is independent of vorticity.

A function called the K-function has been proposed as a measure of vortex motion [17]. It is not clear whether it is, indeed, a local measure of the physical phenomenon known as an eddy or vortex motion. It has been shown that it is a measure of curvature and that it does have significance in fluid dynamics. Requirements laid down for the function were that

1. It is zero for any straight translational motion, but is nonzero for any motion with rotation.
2. It is zero for an irrotational vortex, for rigid rotation, and for the Hagan–Poiseuille flow in a straight conduit.

This arbitrarily defined function is expressed in rectangular Cartesian coordinates as

$$K = \left[\left(\frac{\partial w}{\partial z}\right)\left(\frac{\partial v}{\partial y}\right) - \left(\frac{\partial w}{\partial y}\right)\left(\frac{\partial v}{\partial z}\right)\right] + \left[\left(\frac{\partial u}{\partial x}\right)\left(\frac{\partial w}{\partial z}\right) - \left(\frac{\partial u}{\partial z}\right)\left(\frac{\partial w}{\partial x}\right)\right]$$

$$+ \left[\left(\frac{\partial v}{\partial y}\right)\left(\frac{\partial u}{\partial x}\right) - \left(\frac{\partial v}{\partial x}\right)\left(\frac{\partial u}{\partial y}\right)\right].$$

$$(2.55)$$

Fig. 2.8 Goertler vortices

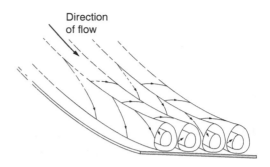

It has been shown that the rate of dissipation is a linear combination of vorticity and the K-function:

$$\Phi = \mu[\Omega - 4K].\tag{2.56}$$

where μ is the fluid viscosity.

This equation indicates that a flow for which K is zero (the curvature is zero) has dissipation proportional to its vorticity. It also shows that a real fluid in motion may be dissipating energy in spite of the absence of vorticity. More importantly, it implies that if a flow has vorticity and has no dissipation, it must also have curvature.

Goertler [18–20] attached considerable significance to the curved flow condition shown in Fig. 2.8. He concluded that the concavity of the wall stabilizes the flow and convexity of the wall destabilizes the flow, and that the critical condition is that the streamlines be concave in the direction of increasing velocity. When such a condition exists, he predicted that longitudinal vortices would form [21]. By using the x–y coordinate system shown in Fig. 2.8 and assuming $\partial v/\partial y$ is zero, it is seen that the K-function is another way of expressing this condition.

Goertler in effect requires that the K-function be negative. If $\partial u/\partial y$ is zero, the K-function is zero; if the streamlines are not curved, the K-function is zero. If the streamlines are convex in the direction of increasing velocity gradient, the K-function is positive. The K-function appears to provide a mathematical measure of the Goertler criterion.

2.12 Generation of Vorticity in a Viscous Boundary Layer: Precursor to Turbulence

A boundary flow will now be considered with the object of elucidating the way in which the vorticity associated with a velocity gradient can be changed into distinct vortices (eddies), as an initial step in the transition to turbulence. The steady two-dimensional flow of an incompressible fluid over a cylinder, neglecting gravity, will serve as the specific flow. A qualitative approach will be adopted. It will be initially

Fig.2.9 Vorticity in flow
over a cylinder

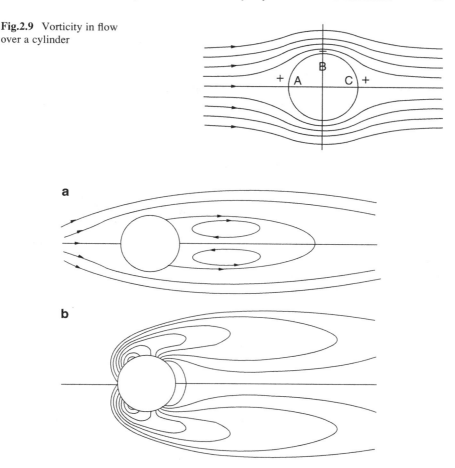

Fig. 2.10 Eddies shed from the cylinder

assumed that the Reynolds number, defined by $Re = \rho RU/\mu$, where R is the cylinder radius and U is the free stream velocity, is small. The no-slip condition results in a boundary layer at the cylinder surface (Fig. 2.9). Flow outside this boundary layer is effectively inviscid since velocity gradients vanish. If U is small, so that velocity gradients within the boundary layer are small, the flow here is also approximately inviscid.

Streamlines around the cylinder are compressed. Generally, they converge between A and B and diverge between B and C. As the fluid is approximately inviscid, the fluid pressure along a streamline decreases between A and B and increases between B and C. The negative pressure gradient between A and B is transformed to kinetic energy and the flow accelerates; with a positive pressure gradient between B and C, the flow decelerates. A constant total pressure (static plus dynamic) along the streamline is maintained. As Re increases, streamlines are increasingly compressed and velocity gradients in the boundary layer become larger. Viscous resistance to shear within the layer becomes significant. Energy is thereby extracted from the as heat. As a consequence, the fluid in the boundary layer arriving at B does not have sufficient kinetic energy to overcome the adverse pressure gradient between B and C, so that fluid close to the cylinder stalls at some point between B and C. The adverse pressure gradient at this point can induce reverse flow close to the cylinder and provide the onset of concentrated vorticity in the lee of the cylinder (Fig. 2.10). The inviscid part of the flow is displaced outward. Thus, the flow is said to "separate" around the site of concentrated vorticity. The shear associated with this separation increases the rotational motion of the fluid next to the cylinder, and this region may become a distinct entity of rotating fluid – an eddy. The eddy, through shear with the surrounding fluid, will further extract energy from the main flow.

The eddies are shed from the cylinder and move downstream with the main flow. Shedding occurs with a regular frequency at low-to-moderate Re values. Eddies enlarge and then are shed alternately from either side of the cylinder. This pattern of regularly spaced eddies moving downstream is referred to as a Von Karman vortex street. This pattern of flow is a precursor to turbulence insofar that with further increase in Re, shedding becomes irregular, the eddy wake becomes disorganized, and a complex velocity field emerges.

2.13 Typical Vorticity Distributions

The first example will be the external flow over an airfoil (see [14]). It will be assumed that the Reynolds number is large and the flow is two dimensional (so that the vorticity vector is always perpendicular to the velocity) – see Fig. 2.11.

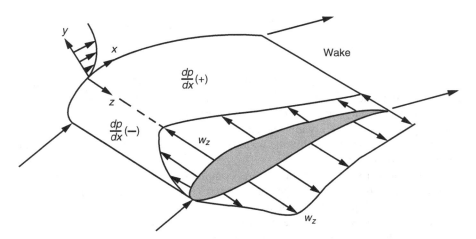

Fig. 2.11 Vorticity distribution in flow over an airflow

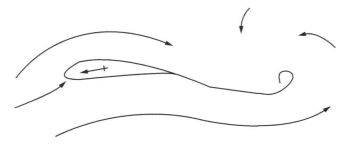

Fig. 2.12 Starting vortex

A local coordinate system with $y = 0$ on the surface of the airfoil and x in the flow direction is chosen. Vorticity diffusion will be primarily normal to the wall. A, the stagnation point so that the positive x-axis is on the upper surface. Curvature will be ignored in this qualitative treatment. The stagnation point is a point of zero shear and hence zero vorticity ($F_{x \text{ viscous}} = \mu \, \partial u / \partial y = -\mu \omega_z$, see [14], where F is the shear stress). As the flow accelerates away from the stagnation point on the upper surface, the shear stress becomes positive and the vorticity becomes negative. In this region, the pressure drops and there is a flux of negative vorticity away from the wall: $\mu \sigma_z = -\mu \partial \omega_z / \partial y = \partial p / \partial x < 0$ where σ_z is the vorticity flux in the z direction. The surface is a source of negative vorticity. Near the front of the airfoil, the pressure reaches a minimum and then slowly increases as the trailing edge is approached. In this region, $\partial p / \partial x$ is positive and the wall in effect absorbs negative vorticity from the flow. The wall flux is positive–negative vorticity diffuses toward the wall. The maximum vorticity now occurs within the flow, as the sign of $\partial \omega_z / \partial y$ is negative at the wall. The process continues until the trailing edge is reached.

On the underside of the airfoil, similar processes occur, but the x coordinate now decreases in the flow direction and the signs of the events change. The pressure gradient that accelerates the flow generates positive vorticity, while the decelerating pressure gradient absorbs positive vorticity. At the trailing edge, the upper and lower streams merge. There is a discontinuity at this point that is washed out as the flow proceeds downstream. The negative vorticity from the upper surface and the positive from the lower merge into the wake. These regions in merging destroy the wake. Assuming vorticity has not diffused very far from the surface at the trailing edge, one can show that the net vorticity across the wake is zero:

$$\text{mean } \omega_z \text{ at trailing edge} = \int_{\delta l}^{\delta_u} \omega_z \mathrm{d}y = \int_{\delta l}^{\delta_u} \frac{\partial u}{\partial y} = (u)_{\delta l}^{\delta_u} = 0.$$

It can also be shown that [14] the net flux of vorticity from the surface of the airfoil is zero. However, there is a net vorticity within the flow. Integrating the vorticity in the region outside the airfoil to a radius R and then letting $R \to \infty$, it is found that, $\int \omega_z \mathrm{d}A = \Gamma$ – a finite number equal to the circulation. The net nonzero

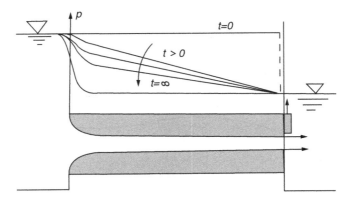

Fig. 2.13 Vorticity generation in channel flow

vorticity is inserted into the flow during the transient process by which the flow is established. In the transient process, the flow does not leave the trailing edge smoothly and the "starting vortex" is formed. Figure 2.12 illustrates a starting vortex formed by impulsively moving the airfoil.

The starting vortex contains the same net amount of vorticity as the airfoil but with opposite sign. A circulation loop around the airfoil and including the starting vortex has $\Gamma = 0$.

The second example will be that of flow through a channel connecting two reservoirs at different elevations – see Fig. 2.13. Consider that initially the fluid is at rest as the exit from the channel is sealed off.

The pressure in the channel is uniform and high. If the seal is rapidly removed, a pressure wave passes through the channel at the speed of sound – much higher than fluid velocities. Instantaneously, a linear pressure gradient is set up in the channel. The stationary fluid has $\omega = 0$ and the pressure forces do not create vorticity. The initial pressure gradient is constant and is required to accelerate the fluid. The momentum equation is

$$\rho \frac{\partial u_x}{\partial t} = -\frac{\partial p}{\partial x} - \mu \frac{\partial \omega_z}{\partial y} = -\frac{\partial p}{\partial x} - \mu \sigma_z. \tag{2.57}$$

The viscous force has been expressed in this equation as a flux of vorticity. The initial vorticity is zero, but there is a flux of vorticity at each wall:

$$-\frac{\partial p}{\partial x} = \mu \sigma_z|_0, \tag{2.58}$$

where $\sigma_z = \partial \omega_z / \partial y$.

The final state will depend on the competing pressure and viscous effects.

Viscous timescale, t_{vis}: this is the time taken for vorticity to diffuse halfway across the channel of width d. The Rayleigh analogy will be used, so that $t_{vis} = (2d)^2/(3.6u_A)$ where v is the kinematic viscosity of the fluid (see [14], p. 347). For $d = 10$ cm and an airflow, t_{vis} is about 60 s (kinematic viscosity = 0.15 cm^2/s). For a viscous vegetable oil, $v = 1.1$ cm^2/s and $t_{vis} = 7$ s.

Suppose the flow is very fast, so that $Re \to \infty$ with d/L finite. In this situation, the vorticity flux term in (2.57) is zero except near the walls. Most fluid particles traverse the channel so fast that vorticity diffusion has no effect on them.

Particles start with no vorticity in the upstream reservoir and traverse the channel in irrotational flow. Vorticity is confined to a small region near each wall. After the pressure gradient generates new vorticity it diffuses only a short distance from the wall before convection moves it downstream and into the exit reservoir.

2.14 Vorticity in a Compressible Fluid

There are several unique features of vorticity in a compressible fluid. The first is that a vortex may develop a vacuum in the core if the strength is enough to centrifuge fluid away from the center of the vortex. Another is vorticity enhancement by fluid compression, represented by the second term in the vorticity equation (2.19), $\omega(\nabla \cdot \mathbf{u})$.

A third unique feature was revealed by Crocco's work [22]. Starting with the momentum equation (2.18) for an inviscid fluid without body forces,

$$\frac{\partial u}{\partial t} + \nabla\left(\frac{\mathbf{u} \cdot \mathbf{u}}{2}\right) - \mathbf{u} \times \omega = \left(\frac{1}{\rho}\right)\nabla p. \tag{2.59}$$

Using the Gibbs equation [23] for the entropy, density and pressure may be eliminated from this equation:

$$T \mathrm{d}s = \mathrm{d}h - \left(\frac{1}{\rho}\right)\mathrm{d}p, \tag{2.60}$$

where T is the temperature, h is the specific enthalpy, and s is the specific entropy. Equation (2.60) can be written as

$$T \nabla s = \nabla h - \left(\frac{1}{\rho}\right)\nabla p. \tag{2.61}$$

Substituting (2.61) in (2.59) gives Crocco's equation:

$$\mathbf{u} \times \omega + T\nabla s = \nabla h_0 + \frac{\partial \mathbf{u}}{\partial t}, \tag{2.62}$$

where h_0 is the stagnation enthalpy, $h_0 = h + 0.5\mathbf{u \cdot u}$.

It has been shown that [24] for steady adiabatic flow of an inviscid without body forces, h_0 is constant along streamlines. The vector ∇h_0 is, therefore, perpendicular to the streamlines, as is the vector $\mathbf{u} \times \boldsymbol{\omega}$ (first and third terms in (2.62)). The term $T\nabla s$ must therefore also be perpendicular to the streamlines. If it is further stipulated that all fluids originate from a constant stagnation enthalpy region, so that $\nabla h_0 = 0$, then $\mathbf{u} \times \boldsymbol{\omega}$ and $T\nabla s$ must be collinear vectors and (2.62) becomes

$$|\mathbf{u}||\omega| + \frac{T\mathrm{d}s}{\mathrm{d}n} = 0, \tag{2.63}$$

where n is perpendicular to the streamlines, in the direction of $\mathbf{u} \times \boldsymbol{\omega}$. This shows that if s is constant, $|\omega|$ must be zero. In other words, *isentropic flows are irrotational* (for steady inviscid flows without body forces and with constant h_0). Thus, if such a specific flow is also isentropic, the well-developed theory of irrotational flow may be used to determine the flow field. If the assumption of isentropicity cannot be made, generally closed form solutions cannot be found and numerical methods must be resorted to.

2.15 Vorticity and the Electromagnetic Analogy

There is a close analogy between [25] calculations involving vortices (in a nonviscous fluid, or superfluid) and calculations involving current filaments in a magnetic field.

This analogy is of particular importance in astrophysical hydrodynamics.

In electromagnetism (mks units),

$$\nabla \cdot \mathbf{B} = 0, \quad \oint \mathbf{B} \cdot \mathbf{dl} = \mu I, \quad W_\mathrm{B} = \frac{B^2}{2\mu}, \tag{2.64}$$

where $\mathbf{B}$ is the magnetic field intensity, $\delta\mathbf{l}$ is an element of a filament, I is the current in the filament, μ is the permeability of the surrounding medium, and W_B is the magnetic field energy density. Now the hydrodynamic equations for a vortex field are as follows:

$$\nabla \cdot \mathbf{v} = 0, \quad \oint \mathbf{v} \cdot \mathbf{dl} = \Gamma, \quad W = \frac{\rho v^2}{2}, \tag{2.65}$$

where $\mathbf{v}$ is the velocity vector, Γ is the vortex strength(circulation), ρ is the fluid density, and W is the velocity field energy density. Equations (2.64) and (2.65) are equivalent with the substitutions:

$$\frac{\mathbf{B}}{\mu} \to \mathbf{v}, \quad I \to \Gamma, \quad \mu \to \rho. \tag{2.66}$$

Recapping, the evolution equation for vorticity was

$$\frac{\partial \omega}{\partial t} = \nu \nabla^2 \omega + \nabla \times (\mathbf{v} \times \omega), \tag{2.67}$$

where $\omega = \nabla \times \mathbf{v}$ and ν is the kinematic viscosity.

Note that $\nabla \cdot \omega = 0$ analogous to the magnetic flux condition $- \nabla \cdot \mathbf{B} = 0$. Batchelor has argued that magnetic fields can be generated by vortex-dominated turbulence. There is then a "magnetic viscosity" η, analogous to the fluid viscosity, leading to a new dimensionless number – the *magnetic Reynolds number* – defined by

$$Re_M \equiv \frac{Ul}{\eta}.$$

As vorticity leads to dissipative processes, in the case of turbulence one would expect that magnetic dissipation will be larger than the ohmic rate. This was first discussed by Spitzer [26] in connection with the magnetic fields of stellar interiors.

It turns out that a frequently encountered field in astrophysical plasmas is the force-free field. The solar corona is one such example. The magnetic helicity is defined as

$$H_M \equiv \int \mathbf{B} \cdot \mathbf{A} dV, \tag{2.68}$$

where $\mathbf{A}$ is the vector potential, and $\mathbf{B} = \nabla \times \mathbf{A}$ (cf. $\omega = \nabla \times \mathbf{v}$). The magnetic helicity is a very important entity in that it is minimal for a force-free field. Just like vorticity, it can be used to determine the field.

The equation for the field evolution with magnetic viscosity is given by

$$\frac{\partial \mathbf{B}}{\partial t} = \nabla \times (\mathbf{v} \times \mathbf{B}) + \eta \nabla^2 \mathbf{B}. \tag{2.69}$$

For magnetic fields, the last term is the same as $-\eta \nabla \times (\nabla \times \mathbf{B})$.

An equation for the vector potential $\mathbf{A}$ is obtained by removing the curl from this equation:

$$\frac{\partial \mathbf{A}}{\partial t} = \mathbf{v} \times \mathbf{B} - \nabla \Phi - \eta \nabla \times \mathbf{B}. \tag{2.70}$$

The scalar products of (2.69) with $\mathbf{A}$ and of (2.70) with $\mathbf{B}$ are taken, added, and the remaining terms integrated over volume. This leads to an evolution equation for the magnetic helicity:

$$\frac{\partial}{\partial t} \int \mathbf{A} \cdot \mathbf{B} d\mathbf{x} = -\eta \int (\nabla \times \mathbf{A}) \cdot (\nabla \times \mathbf{B}) d\mathbf{x}$$
$$- \eta \int (\nabla \times \mathbf{B}) \cdot \mathbf{B} d\mathbf{x} = -2\eta \left(\frac{4\pi}{c}\right) \int \mathbf{J} \cdot \mathbf{B} d\mathbf{x}, \tag{2.71}$$

where $\mathbf{B} = \nabla \times \mathbf{A}$ and $\nabla \times \mathbf{B} = (4\pi/c)\mathbf{J}$.

This equation states that the evolution of magnetic helicity in a fluid is driven by dissipation and that if $\mathbf{B}$ is parallel to $\mathbf{J}$ the magnetic field decays; otherwise, it increases. Also, in a fluid with $\eta \to 0$, the magnetic field is a conserved quantity. Just as in the case of vorticity in a fluid, the magnetic helicity serves as a topological tool for understanding the dissipative mechanisms and the instabilities of the fluid. Magnetic field lines tangle, merge, split, and eventually decay.

2.16   Quantization of Circulation and Vorticity

The description of a superfluid in terms of a single wave function, $|\psi| \exp(iS)$, where S is the phase function, leads to the following superfluid current equations:

$$j_s = \left(\frac{h}{2\pi}\right)|\psi|^2\nabla S \quad \text{(HeII, superfluid)}, \tag{2.72}$$

$$\mathbf{J}_e = \left(\frac{eh}{2\pi m}\right)|\psi|^2\nabla S - \left(\frac{2e^2}{m}\right)|\psi|^2\mathbf{A}. \tag{2.73}$$

The mass current density j_s is used for He II and the electric current density $\mathbf{J}_e$ for the superconductor (valid for cubic superconductors only [27]).

Consider He II in an annular region between two concentric cylinders (a multiply connected region), as depicted in Fig. 2.14.

The temperature is absolute zero, so that the He II is a pure superfluid.

To determine the flow pattern, consider the circulation

$$\Gamma = \oint v_s \cdot dl, \tag{2.74}$$

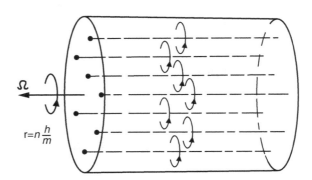

Fig. 2.14 Helium II between two concentric cylinders

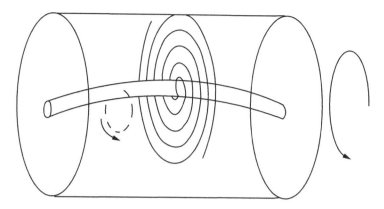

Fig. 2.15 Rotating drum with superfluid in it

where v_s is the velocity of the superfluid and the integral is taken around any contour wholly within the fluid. Equation (2.72) implies [27] and the superfluid equation can be written as

$$v_s = \left(\frac{h}{2\pi m_4}\right)\nabla S, \qquad (2.75)$$

where m_4 is the atomic mass of the ^{4}He atom, and so it is possible to express the circulation in terms of the wave function phase S:

$$\Gamma = \left(\frac{h}{2\pi m_4}\right)\oint \nabla S \cdot dl. \qquad (2.76)$$

For the circle 1 in Fig. 2.14, the circulation is:

$$\Gamma = \left(\frac{h}{2\pi m_4}\right)(\Delta S)_1. \qquad (2.77)$$

Since the superfluid is single valued, a traverse around a closed contour must leave it unaltered, so that the change in S can be only an integral multiple of 2π or zero. From (2.77) it can be seen that the circulation (and hence the vorticity) is quantized, with the values

$$\Gamma = n\left(\frac{h}{m_4}\right) \quad \text{where } n = 0, 1, 2, \qquad (2.78)$$

h/m_4 is known as the quantum of circulation and has the value 9.98×10^{-8} m^2/s.

Vinen [28] was the first to experimentally demonstrate that circulation is quantized in He II. The apparatus is sketched in Fig. 2.15 and consisted of a cylindrical

Fig. 2.16 The Magnus force
is the perpendicular lift force
on the wire when it and the
vortex are dragger at
velocity V

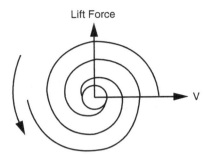

container with a fine conducting wire stretched along its axis. The He II filling the
container thus occupied a multiply connected region.

When fluid is encircling the wire and the wire is vibrating, the Magnus force induces
circular motion in the wire. The circulating fluid is dragged along with the wire

A solid cylinder, around which there is a fluid circulation, experiences a force –
the Magnus force – when it moves through the fluid (see Fig. 2.16).

Treating the superfluid as an ideal Euler fluid, the Magnus force on the wire in
Vinen's experiment is given by

$$f_M = \rho_s \Gamma \times \mathbf{V}, \tag{2.79}$$

where $\mathbf{V}$ is the wire velocity relative to the superfluid outside the circulation and Γ
is the vector indicating the sense and strength of the circulation.

The wire was placed in a magnetic field and transverse vibrations excited by
passage of an AC electric current through it. When the surrounding liquid is not
rotating, the normal modes of the wire are two plane polarized waves at right angles
to each other and with the same frequency. When the superfluid circulates smoothly
around the wire, the latter is acted upon by the Magnus force, causing both planes of
vibration to process. The normal modes of the wire can now be viewed as circularly
polarized in opposite directions, with frequencies differing by $\Delta v = \rho_s \Gamma / 2\pi W$
where W is the sum of the mass per unit length of the wire plus half the mass of
the fluid displaced by this length. The difference Δv appears as a beat frequency of
the voltage induced in the wire and this provided a direct method of measuring the
circulation Γ.

2.17 Quantized Vortices in He II

Rotation of the superfluid: when the two-fluid model of He II was first suggested, it
was believed that it would be difficult to set the superfluid fraction into rotation
because superfluid flow was characterized by the irrotationality condition introduced
by Landau [29]:

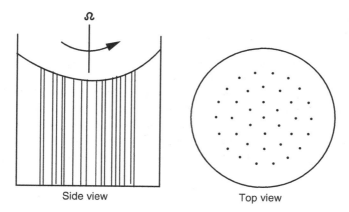

Side view   Top view

Fig. 2.17 Array of quantized vortices in a rotating bucket. Some vortices are missing near the outer edge

$$\nabla \times \mathbf{v}_s = 0. \qquad (2.80)$$

Consider He II contained in a cylindrical bucket and the contour 2 – see Fig. 2.17.

By Stokes' theorem, the circulation around the contour 2 can be written as an integral over the surface **A** enclosed by the contour:

$$\Gamma = \oint {}_2\mathbf{v}_s \cdot d\mathbf{l} = \int {}_A(\nabla \times \mathbf{v}_s) \cdot d\mathbf{A}. \qquad (2.81)$$

Combining (2.80) and (2.81),

$$\Gamma = \oint {}_2\mathbf{v}_s \cdot d\mathbf{l} = 0 \qquad (2.82)$$

Thus, the circulation for any contour in the continuous fluid is zero. If (2.82) holds everywhere in the superfluid, rotation is not possible. But it only holds if $\mathbf{v}_s$ is zero at every point. This state of He II from which superfluid rotation is completely absent is called the Landau state. The rotation of the superfluid can be explained by assuming that it is threaded by parallel straight vortex lines – see Fig. 2.17. It has been explained above how one can have a circulation around a region from which the superfluid is excluded. Contours that enclose a solid obstacle or a vortex core (cf. Contour 1 in Fig. 2.14) yield a quantized circulation:

$$\Gamma = \oint {}_1\mathbf{v}_s \cdot d\mathbf{l} = \oint {}_A(\nabla \times \mathbf{v}_s) \times d\mathbf{l} = \frac{nh}{m_4}. \qquad (2.83)$$

Note: the core of the vortex is defined by that region in which curl $\mathbf{v}_s \neq 0$.

The earliest suggestion that vorticity should play an important role in superfluid hydrodynamics is credited to Onsager [30].

Vortex lines in rotating He II:

It can now be explained how, in Osborne's [31] rotating bucket experiment, the presence of a uniform array of vortex lines enables the superfluid to undergo solid-body rotation. Suppose there are n_v vortex lines per unit area of the bucket, all with their cores parallel to the axis of rotation (Fig. 2.17) and each with the same circulation Γ, represented by an axial vector of magnitude Γ pointing in the direction consistent with the sense of rotation. The strength of the array is specified by the vorticity ϖ, defined to be equal to the total circulation within unit area

$$\boldsymbol{\varpi} = n_v \Gamma. \tag{2.84}$$

From (2.81) it can be seen that ϖ can be identified with the average value of curl $\mathbf{v}_s$. When curl $\mathbf{v}_s$ is nonzero it indicates the presence of vortices. In the bucket the total circulation enclosed by a contour of radius R centered on the axis is $\pi n_v R^2 \Gamma$. For the superfluid to appear to rotate with uniform angular velocity Ω, the total circulation must also be equal to $2\pi R(R\Omega)$. Thus, the condition for the simulation of solid-body rotation is that the vortex-line density is

$$n_v = \frac{2\Omega}{\Gamma}. \tag{2.85}$$

Alternatively, this can be used to determine the required vorticity,

$$\boldsymbol{\varpi} = \nabla \times \mathbf{v}_s = 2\Omega. \tag{2.86}$$

Hall [32] has shown that the ground state of He II contains a regular array of vortex lines all having the smallest possible circulation h/m_4 and with a maximum total number of lines. Viewed from a frame rotating with the container, the equilibrium vortex array is a triangular lattice. It can be shown [27, p. 186] that the critical angular velocity for the formation of one vortex with minimum circulation is

$$\Omega_{c1} = \left(\frac{h}{2\pi m_4 R_0^2} \right) \ln \left(\frac{R_0}{a_0} \right) \tag{2.87}$$

For a typical value of R_0 of 1 cm, $\Omega_{c1} \approx 10^{-3}$ rad/s. Thus, it is easy for vortex lines to appear.

For sufficiently low angular velocities of the container ($\Omega \leq \Omega_{c1}$), the equilibrium state of the superfluid is the Landau state as described earlier. The Landau state was observed by Hess and Fairbank [33] in an experiment in which liquid 4He was cooled through the λ point while inside a rotating cylinder made from a closed capillary tube. Above T_λ, the He I was in solid-body rotation. On starting from rest, provided the cylinder was rotated slowly enough, it was found that the superfluid formed in a state of zero angular momentum relative to the laboratory. Packard and

Sanders [34] have developed a method of counting vortex lines that exploits the trapping of electrons on vortex cores. This was used to demonstrate that a small amount of vorticity could be detected for up to 30 min after the container was brought to rest. This highlighted an important aspect of vorticity: the *persistence of vorticity*.

A single electron self-trapped inside a cavity whose size is fixed by a balance between the outward pressure due to the zero-point motion of the electron and the inward pressure of the surrounding liquid is known as a negative ion, and can be trapped on the cores of vortex lines. The positive ion, which can also be trapped on vortex lines, is a "snowball" consisting of an α particle to which several neutral ^{4}He atoms are bound electrostatically. It has been shown that [35] the motion of ions in He II can create quantized vortex rings, which become coupled to the ions. The mechanism by which a bare ion nucleates a vortex ring in He II and then becomes trapped on it is unknown. The solution of this problem will probably cast light on the creation of vorticity on a microscopic scale. This knowledge is required in order to understand the breakdown of superfluidity through the agency of vorticity.

References

1. Truesdell, C.: The Kinematics of Vorticity. Indiana University Press, Bloomington (1954)
2. Thompson, W. (Lord Kelvin): On vortex motion. Math. Phys. 4, 49 (1869)
3. Batchelor, G.K.: An Introduction to Fluid Dynamics. Cambridge University Press, Cambridge (1967)
4. Hadamard, J.: Sur les Tourbillons Produit par les Ondes De Choc. Note III. Lecons sur la Propagation des Ondes. In: Herman, A. (ed.). Paris, 362 (1903)
5. Vazsonyi, A.: On rotational gas flows. Quart. J. Appl. Math. 3(1), 29–37 (1945)
6. Holton, J.R.: An Introduction to Dynamic Meteorology. Academic, New York, NY (1979)
7. Green, S.I. (ed.): Fluid Vortices. Kluwer Academic, Boston, Massachusetts (1995) (Chapter 11)
8. Oseen, C.W.: Uber Wirbelbewegung in einer Reibenden Flussigkeit. Ark. f. Mat. Astron. Fyz 7, 14 (1912)
9. Lighthill, M.J.: Boundary layer theory. Part II. In: Rosenhead, L. (ed.) Laminar Boundary Layers, p. 389. Dover, Mineola (1963)
10. Morton, B.R.: The generation and decay of vorticity. Geophys. Astrophys. Fluid Dyn. **28**, 277–308 (1984)
11. Heimenz, H.: On integrals of the hydrodynamical equations which express vortex motion. Dingler's Polytech. **326**, 311 (1867)
12. Burgers, J.M.: A mathematical model illustrating the theory of turbulence. Adv. Appl. Mech. **17**, 1 (1948)
13. Hill, M.J.M.: On a spherical vortex. Phil. Trans. Roy. Soc. Lond. A **185**, 213 (1894)
14. Panton, R.I.: Incompressible Flow. Wiley, New York, NY (1984). p. 333
15. Bjerknes, V.: Uberdie Bildung Circulations bewegung Skrifter (1898)
16. Lamb, H.: Hydrodynamics. Cambridge University Press, Cambridge (1932)
17. Mockros, L.: Ph.D. Thesis, University of California, Los Angeles (1962)
18. Goertler, H.: Uber eine Dreidimensionale Instabilitat Laminaren Grenzschichten am Koncaven Wander. Naschr Wiss Gas,Gottingen Math Phys Klasse, 2(1) (1940)
19. Witting, H.: Theorie der Sekundaren Instabilitat der Laminaren Grenzschichten. In: Goertler, H. (ed.) Boundary Layer Research. Springer, Berlin (1958)

20. Witting, H.: Einige Neuer Ergebnisse zur Hydrodynamischen Stabilitats Theorie. Zeits.fur Flugwissenschaften 8 Jahrgang, Heft 1 (1960)
21. McCormack, P., Welcher, H.: J. Heat Transfer (1967)
22. Crocco, L.: Zeits. Angew. Math. Mech. **17**, p. 1 (1937)
23. Black, W.Z., Hartley, J.G.: Thermodynamics. Harper Collins, New York, NY (1991)
24. Currie, I.G.: Fundamental Mechanics of Fluids. McGraw-Hill, New York (1974)
25. Fetter, A., Donnelly, R.: On the equivalence of vortices and current filaments. Phys. Fluids **9**, 619 (1966)
26. Spitzer Jr., L.: Influence of fluid motions in the decay of an external magnetic field. Astrophys. J. **125**, 525 (1957)
27. Tilley, D.R., Tilley, J.: Superfluidity and Super-Conductivity, 3rd edn. Graduate Student Series in Physics, Institute of Physics Publishing, Philedelphia (1990)
28. Vinen, W.F., et al.: Quantized Vortex Dynamics and Superfluid Turbulence. Springer, Berlin (2001)
29. Landau, L.D.: The theory of superfluidity of helium II. J. Phys. Moscow **5**, 71 (1941)
30. Onsager, L.: Nuevo Cimento. **6**(2), 249 (1949)
31. Osborne, L.: Proc. Phys. Soc. **A63**, 909 (1949)
32. Hall, H.E.: On the rotation of liquid helium II. Adv. Phys. **9**, 89 (1960)
33. Hess, G.B., Fairbank, W.M.: Measurements of angular momentum in superfluid helium. Phys. Rev. Letts. **19**, 216 (1967)
34. Packard, R.F., Sanders, T.M.: Observation of single vortex lines in rotating superfluid helium. Phys. Rev. **A6**, 799 (1972)
35. Rayfield, G.W., Reif, F.: Quantized vortex rings in superfluid helium. Phys. Rev. **136**, 1194 (1964)

Chapter 3
The Nanoboundary Layer and Nanovortex Core

3.1 Introduction

In this chapter, nanoscale fluid flow regions with intense vorticity will be considered – (1) the nanoboundary layer within 100 nm of a solid surface over which fluid is flowing and (2) the nanovortex core within 100 nm of the vortex center. In such regions, the kinetic theory of fluids must be used to predict the physical properties of the fluid, and the fluid has intense vorticity, or molecular spin.

3.1.1 Importance of the Nanoscale

One nanometer is 10^{-9} m or 10 Å. It is 10,000 times smaller than the diameter of a human hair. One cubic nanometer is approximately 20 times the volume of a single atom. Figure 3.1 shows various size ranges for different nanoscale objects, starting with small entities such as ions, atoms, and molecules. The size ranges of a number of nanotechnology-related objects (such as quantum dot, nanotube, and single-electron transistor diameters) are also shown in Fig. 3.1. It is obvious that nanoscience, nanoengineering, and nanotechnology all deal with very small objects and systems. Our ability to control and manipulate nanostructures and nanosystems will make it possible to exploit the new physical, biological, and chemical properties of systems that are intermediate in size between single atoms, molecules, and bulk material. Some of the reasons why nanoscale has become so important are:

1. The quantum mechanical (wavelike) properties of electrons inside matter are influenced by variations on the nanoscale.
2. Nanoscale components have very high surface to volume ratio, making them ideal for use in composite materials, reacting systems, drug delivery, and chemical energy storage.
3. Macroscopic systems made up of nanostructures can have much higher density than those made up of microstructures. They can also be better conductors of

P. McCormack, *Vortex, Molecular Spin and Nanovorticity: An Introduction*,
SpringerBriefs in Physics, DOI 10.1007/978-1-4614-0257-2_3,
© Percival McCormack 2012

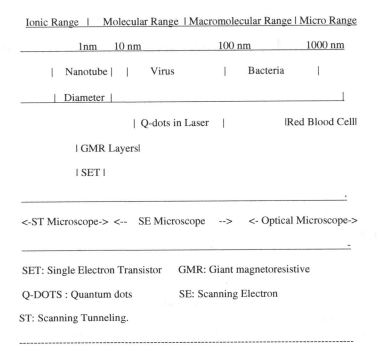

Fig. 3.1 Size ranges for nanoscale objects

heat and electricity. This can result in new electronic device concepts; smaller and faster circuits; and greatly reduced power consumption and energy dissipation.

The vortex core is a flow region characterized by streamline curvature and uniform vorticity. It rotates as a rigid, or solid, body. It will be shown in this paper, that to be self-consistent thermodynamically, the macroscopic rotation must be accompanied by concomitant internal rotation of the molecules in the core fluid. This is true for two-dimensional vortices (rectilinear) and three-dimensional vortices (ring). It will also be true for the helical vortices in the shear layer at fluid/wall interfaces.

Solid-body rotation implies that although there is vorticity in the core, there is no shear viscosity (or internal energy dissipation). The surrounding irrotational fluid has no vorticity, but has shear viscosity. Through interaction with the boundary of the core, energy dissipation occurs there.

This produces a region in which vorticity and dissipation are present. The layer grows deeper with time and the core size is correspondingly reduced. The discontinuity in the first derivative of the fluid velocity is thereby removed. At any point within a volume V of a continuum, it can be shown that

$$\varepsilon_{ijk}\sigma_{kj} = 0, \tag{3.1}$$

where ε_{ijk} is the permutation symbol and σ_{kj} is an element of the stress tensor at that point. This implies that the stress tensor is symmetrical, i.e., $\sigma_{ij} = \sigma_{ji}$.

It is true, however, only in the absence of a "body couple" of order V (for example, the couple exerted on a polarized dielectric medium by an external electric field).

The torque on a fluid element exerted by the surrounding fluid is proportional to the antisymmetric part of the stress tensor and remains finite as the fluid element is reduced to a point. To avoid the nonphysical situation of a finite force (or moment thereof) acting on an element of infinitesimal inertia, one must conclude that the antisymmetric part of the tensor vanishes identically. This limiting operation, however, has no physical significance. The concept of a torque acting on a point is meaningless. Thus, a continuum mechanics that excludes body couples cannot be expected to determine the physical properties of the vortex core. It does predict that a fluid flow, in which there is vorticity and no energy dissipation, must be curved.

Molecular kinetic theory (nanofluidics) and thermodynamics are not limited in this way. It can be shown that the thermodynamic internal energy E, a good physical variable, is a function of (among other parameters) the difference between the rate of angular displacement of the molecule (θ) and half the curl of the rate of positional displacement (**S**):

$$E = E\left(\theta - \left(\frac{1}{2}\right)\nabla \times \mathbf{S}\right). \tag{3.2}$$

Curl **S** corresponds to a uniform solid-body rotation of the system. In a diatomic system $(1/2)$ curl **S** corresponds to a uniform rotation of the molecular mass centers. But E must be invariant under this transformation of coordinates. For a diatomic medium, it is only when $\theta - (1/2)\nabla \times \mathbf{S} = 0$ that the rotation of the mass centers is balanced out by a concomitant rotation of the molecular axes. It is only for a motion of this kind that E is independent of $\nabla \times \mathbf{S}$. Thermodynamics requires, then, that there must be a direct relation between macroscopic (local) fluid rotation and molecular rotation (or spin).

It will be shown in this paper that the kinetic theory of a dilute gas (two particle collisions only) of rotating molecules in a vortex core (with matching macroscopic, or local, rotation) predicts a polarization of the internal molecular rotations about the local rotation axis – a Barnett effect [1]. In such a spin-aligned system, or nanosystem, changes in the physical properties of the fluid result. Electric polarization, coefficient of heat conduction, and optical properties of the fluid in the vortex core will be considered.

In 1971, the author [2] measured the flame speed in vortex rings formed from a premixed composition of propane and air. The average flame speed was determined to be 300 cm/s – about ten times the laminar flame for the mixture. This can be explained by an enhanced coefficient of heat conduction in the vortex core, as will be established here.

3.2 Kinetic Theory of the Vortex Core Gas

The gas in a vortex core then consists of rotating molecules (in the classical sense). Waldman [3, 4] has derived a kinetic theory for a dilute gas (two particle collisions only) of rotating molecules. In the presence of irreversible processes such as diffusion or heat conduction, partial polarization (or alignment) of the molecular axes can be shown to occur. By an extension of this theory, it can be shown that in the presence of local rotation (macroscopic) polarization of the molecular rotational axes occurs along the local rotation axis ([1]: Barnett polarization). A brief review of Waldman's theory will now be presented.

A dilute gas of molecules that can translate and rotate is considered. The dynamic state of such a gas is described by the one-particle distribution function:

$$f = f(t, \mathbf{r}, \mathbf{v}, \mathbf{w}), \tag{3.3}$$

where $\mathbf{v}$ is the translational velocity vector, $\mathbf{w}$ is the angular velocity of rotation vector, $\mathbf{r}$ is the position vector, and t is the time.

The dynamics of the system (in the absence of an external force) is governed by a Boltzmann transport equation of the type,

$$\frac{\partial f}{\partial t} + \mathbf{v} \cdot \frac{\partial f}{\partial \mathbf{r}} + \omega(f) = 0, \tag{3.4}$$

where $\omega(f) \equiv -(\partial f/\partial t)_{\text{coll}}$ is a linear collision operator.

If f' is the precollisional function and f the postcollisional, then the operator can be shown to have the following form:

$$w(f) = n_s R^2 \int (f - f') \mathbf{v} \cdot \mathbf{k} \mathrm{d}^2 k, \tag{3.5}$$

where n_s is the concentration of molecules, R is the effective scattering radius, and $\mathbf{k}$ is the unit vector lying along the line segment from the center of the scattering circle to the exit point of the molecule. For small deviations from thermal equilibrium, one can write

$$f = f_0(1 + \Phi), \tag{3.6}$$

where f_0 is the equilibrium distribution function. Defining the dimensionless parameters,

$$\mathbf{V} = [\sqrt{(m/2k_BT_0)}]\mathbf{v}, \tag{3.7a}$$

$$\mathbf{W} = [\sqrt{(\Theta/2k_BT_0)}]\mathbf{w}, \tag{3.7b}$$

where k_B is Boltzmann's constant, m is the molecular mass, and Θ is the molecular moment of inertia about the rotation axis, then

$$\Phi = \Phi(\mathbf{V}, \mathbf{W}). \tag{3.8}$$

As a basis set for $\mathbf{V}$, $\mathbf{W}$ space, the eigensolutions for coupled harmonic oscillators are taken. These form the Clebsch–Gordon series and have the form,

$$\Phi_{LMl}^n(\mathbf{V}, \mathbf{W}) = \Sigma_m C(LlMm)\Phi_{lm}^n(\mathbf{V})\Phi_{lm}^n(\mathbf{W}). \tag{3.9}$$

Depending on the values of $N = l + 2n$ and l, these are scalars, vectors or pseudo-vectors, and Cartesian tensors or pseudo-tensors (parity $(-1)^l$ and $(-1)^{l+1}$, respectively). For example,

$N = 0, l = 0$ gives the scalar $\Phi^{(0)} = 1$
$N = 1, l = 1$ gives the vector $\Phi_\mu^{(1)} = (\sqrt{2})V_\mu$

gives the pseudo-vector $\Psi_\mu^{(1)} = (\sqrt{3}/w_0)W_\mu$
where $w_0 = \sqrt{(3k_BT_0/\Theta)}$

$N = 2, l = 2$ gives the tensor $\Phi_{\mu\nu}^{(1)} = [V_\mu V_\nu - (1/3)V^2\delta_{\mu\nu}]$

gives the pseudo-tensor $\Psi_{\mu\nu}^{(1)} = (1/w_0)\langle\langle V_\mu V_\nu\rangle\rangle$

where $\langle\langle\rangle\rangle$ implies the symmetric vanishing trace and $\mu, \nu = 1, 2, 3$ correspond to Cartesian components and are determined by the various products of the unit vectors or tensors $\Phi_{lm}^n(\mathbf{V})$ and $\Phi_{lm}^n(\mathbf{W})$.

The distribution function is expanded in terms of these scalars, vectors, and tensors:

$$f = f(1 + \Phi) = f_0[1 + \Sigma_k(\Phi^{(k)}a^{(k)} + \psi^{(k)}b^{(k)} + a_\mu^{(k)}\Phi_\mu^{(k)} + b_\mu^{(k)}\Psi_\mu^{(k)} + \Psi_{\mu\nu}^{(k)}b_{\mu\nu}^{(k)} + \ldots\ldots)]. \tag{3.10}$$

Multiplying this equation across by $\Phi^{(i)}$, $\Psi^{(i)}$, etc., integrating in $\mathbf{v}$, $\mathbf{w}$ space and using orthogonality relations for mean values of the tensor products, the coefficients $a^{(k)}$, $b^{(k)}$, $a_\mu^{(k)}$, ..., etc. can be evaluated. These can be identified with well-known physical parameters of the system. For example,

$$a^{(1)} = \frac{n - n_0}{n} \Rightarrow \text{variation of particle density}, \tag{3.11}$$

$$b^{(1)} = (\sqrt{3}n)\langle \mathbf{v} \cdot \mathbf{w} \rangle / n_0 v_0 w_0 \Rightarrow \text{longitudinal spin}, \qquad (3.12)$$

$$\mathbf{a}_\mu^{(1)} = (\sqrt{3})\mathbf{j}/n_0 v_0,$$

where $\mathbf{j} = n\langle \mathbf{v} \rangle \Rightarrow$ gas velocity – this is a polar vector, (3.13)

$$\mathbf{b}^{(1)} = (\sqrt{3})n\langle \mathbf{w} \rangle / n_0 w_0 \Rightarrow \text{vector polarization – this is a pseudo - vector}, (3.14)$$

$$\langle\langle \mathbf{a}^{(1)} \rangle\rangle = \frac{3}{\sqrt{2}} \left[\frac{n}{n_0 v_0{}^2} \right] \langle v_\mu v_\nu - \left(\frac{1}{3} \right) v^2 \delta_{\mu\nu} \rangle \Rightarrow \text{friction tensor}, \qquad (3.15)$$

$$\langle\langle \mathbf{a}^{(2)} \rangle\rangle = \frac{\sqrt{15}}{2} \left[\frac{n}{w_0{}^2 n_0} \right] \langle\langle w_m w_n \rangle\rangle \Rightarrow \text{tensor polarization}. \qquad (3.16)$$

3.3   Effect of Local (Macroscopic) Rotation of the Gas

To account for this, a term must be added to the equation of motion for $\langle \mathbf{w} \rangle$, which will give the Barnett effect for local thermal equilibrium. It will be found that the correct antisymmetric part of the pressure tensor now appears in this equation. This term must be added to the pressure tensor in the conservation equation for momentum. No other changes are necessary in the system equations.

Consider the equilibrium distribution function $f_{0\omega}$ of a gas rotating as a rigid body with angular velocity $\boldsymbol{\omega}$,

$$f_{0_\omega} \propto \exp - \beta \left[\left(\frac{1}{2} \right) m \mathbf{v}^2 - (\mathbf{l} + I\omega) \cdot \omega \right], \qquad (3.17)$$

where $\mathbf{v}$ is the particle velocity in the laboratory system, $\mathbf{l} = \mathbf{r} \times m\mathbf{v}$ is the orbital angular momentum of a particle ($\mathbf{r} = 0$ is a point on the axis of local rotation), and $I\boldsymbol{\omega}$ is the internal angular momentum of the particle.

The distribution function is

$$f_{0_\omega} = n_0(r) \left(\frac{m\beta_0}{2\pi} \right) \exp \left\{ -\left(\frac{1}{2} \right) \beta_0 [m(\mathbf{v} - \mathbf{v}_\omega)^2 + I\mathbf{w} \cdot \omega] \right\}, \qquad (3.18)$$

where

$$n_0(r) = \frac{N \exp[(1/2)\beta_0 m v_\omega(r)^2]}{\int d^3 r' \exp[(1/2)\beta_0 m v_\omega(r')^2]}, \qquad (3.19)$$

and N is the total number of particles. Using this distribution function and carrying out the integrations in $\mathbf{v}$, $\mathbf{w}$ space, the following expression for the mean angular velocity of the particles is obtained:

$$\langle \mathbf{w} \rangle_{0\omega} = w_0^2 \beta_0 I \omega = \left(\frac{1}{2}\right) w_0^2 \beta_0 \nabla \times \langle \mathbf{v} \rangle_{0\omega}, \tag{3.20}$$

where

$$2\omega \equiv \nabla \times \langle \mathbf{v} \rangle_{0\omega}.$$

One sees that the particles (or molecules) are aligned along the direction of the axis of local (solid-body) rotation and that there is a unique relation between the mean particle angular velocity and the local angular velocity.

This is a linear approximation and at this level of approximation there will be no alignment of $\langle\langle w_\mu w_\nu \rangle\rangle$ or higher spin tensors, as these would be at least quadratic in $\boldsymbol{\omega}$. Thus, only the transport equations for $\mathbf{a}^{(1)}$ (proportional to the velocity) and $\mathbf{b}^{(1)}$ – the vector polarization or dipole density (proportional to the angular velocity) – will have to be changed.

The transport equation for $\mathbf{b}^{(1)}$ will be considered here.

In the (3.14) for $\mathbf{b}^{(1)}$, the term $\omega_{+1}^{(11)} b_\mu^{(1)}$ is the relaxation term for $\langle \mathbf{w} \rangle$. A Barnett term must be subtracted from this. The term will be of the form

$$\nabla \times \mathbf{v} \text{ or } \nabla \times \mathbf{a}^{(1)} = \varepsilon_{\mu\lambda\nu} \left(\frac{\partial a_\lambda^{(1)}}{\partial x_\nu}\right).$$

For dimensional homogeneity, this must be multiplied by a constant (l_0) with the dimensions of length and the form which is chosen is

$$\left(\frac{1}{2}\right) l_0 \varepsilon_{\mu\lambda\nu} \left[\frac{\partial a_\lambda^{(1)}}{\partial x_\nu}\right].$$

It is easily shown that l_0 is given by

$$l_0 = w_0 v_0 I \beta_0. \tag{3.21}$$

The equation for $\mathbf{b}^{(1)}$ becomes

$$\partial b_\mu^{(1)}/\partial t + (v_0/3)\partial b_\mu^{(1)}/\partial x_\mu - (v_0/\sqrt{6})\varepsilon_{\mu\nu\lambda}\partial a_\lambda^{(2)}/\partial x_\nu + (v_0/3)\partial b_{\mu\nu}^{(1)}/\partial x_\nu$$
$$+ \omega_{+1}^{(11)}[b_\mu^{(1)} - (l_0/2)\varepsilon_{\mu\nu\lambda}\partial a_\lambda^{(1)}/\partial x_\nu] + \Sigma_{k=2}^{3}\omega_{+1}^{(1k)}b_\mu^{(k)} = 0. \tag{3.22}$$

This equation can be written in terms of the mean angular velocity, which (omitting the sum term) is as follows:

$$n_0 I \partial \langle w_\mu \rangle / \partial t + \partial W_{\mu\nu} / \partial x_\nu + n_0 \omega_{+1}^{(11)} [I \langle w_\mu \rangle - (m l_0^2 / 2) \varepsilon_{\mu\nu\lambda} \partial \langle v_\lambda \rangle / \partial x_\nu] = 0. \quad (3.23)$$

$W_{\mu\nu}$ is the spin flux tensor and is given by

$$W_{\mu\nu} = \left(\frac{n_0}{\sqrt{3}} \right) I w_0 v_0 \left[\left(\frac{1}{3} \right) b^{(1)} \delta_{\mu\nu} - \left(\frac{1}{6} \right) \varepsilon_{\mu\nu\lambda} a_\lambda^{(2)} + \left(\frac{1}{\sqrt{3}} \right) b_{\mu\nu}^{(1)} \right]$$

$$= n_0 I \left[\left(\frac{1}{3} \right) \langle \mathbf{v} \cdot \mathbf{w} \rangle \delta_{\mu\nu} - \left(\frac{1}{2} \right) \langle v_\mu w_\nu - v_\nu w_\mu \rangle + \langle \langle v_\mu v_\nu \rangle \rangle \right]$$

$$= n_0 I \langle w_\mu v_\nu \rangle. \quad (3.24)$$

The complete pressure tensor may be written as

$$p_{\mu\nu} = nkT \delta_{\mu\nu} + \{p_{\mu\nu}\} + [p_{\mu\nu}], \quad (3.25)$$

where the first term is the hydrostatic pressure, $\{p_{\mu\nu}\}$ is the antisymmetric part, and $[p_{\mu\nu}]$ is the symmetric part.

Now

$$\langle \langle p_{\mu\nu} \rangle \rangle = \left(\frac{\sqrt{2}}{3} \right) v_0^2 m n a_{\mu\nu}^{(1)}, \quad (3.26)$$

where $a_{\mu\nu}^{(1)}$ is the shear friction tensor. But there is no shear in the vortex core and so this is zero. Thus, there is no contribution to $p_{\mu\nu}$ from the symmetric part. Also, due to the conservation of total angular momentum,

$$\varepsilon_{\mu\nu\lambda} \left(\frac{n_0 I \partial \langle w_\lambda \rangle}{\partial t} + \frac{\partial W_{\mu\nu}}{\partial x_\nu} \right) = 2 \langle \langle p_{\mu\nu} \rangle \rangle = p_{\mu\nu} - p_{\nu\mu}. \quad (3.27)$$

Comparing with (3.23) one sees that

$$\langle \langle p_{\mu\nu} \rangle \rangle = (-1/2) n_0 \omega_{+1}^{(11)} [\varepsilon_{\mu\nu\lambda} I \langle w_\lambda \rangle - m l_0^2 \partial \langle \langle v_\nu \rangle \rangle / \partial x_\mu], \quad (3.28)$$

where

$$\frac{\partial \langle \langle v_\nu \rangle \rangle}{\partial x_\mu} = \left(\frac{1}{2} \right) \left[\frac{\partial \langle v_\nu \rangle}{\partial x_\mu} - \frac{\partial \langle v_\mu \rangle}{\partial x_\nu} \right]. \quad (3.29)$$

Thus, when the polarization equals the local Barnett polarization, $p_{\mu\nu}$ vanishes, and there is no shear or bulk viscosity contribution to the pressure tensor. The only pressure contribution is due to the hydrostatic effect. Steele [5] has also considered the dynamics of a system of rotating Brownian particles (nanoparticles). If ξ_I is the ith element of the diagonalized rotational friction tensor, he showed that

$$\xi_i^2 = \left(\frac{2I}{\pi}\right)\left(\frac{\partial^2 V}{\partial \psi_i^2}\right),$$
(3.30)

where I is the moment of inertia of the particle about the relevant axis, V is the potential energy of the particle, and ψ_I is the orientational angle.

If a particle rotates about an axis of symmetry, then V is independent of the angle of orientation about this axis and the corresponding friction tensor component is zero. Thus, in a gas of particles rotating about symmetry axes, which are aligned with one another, there will be no relaxation of the rotational angular momenta of the molecules. The relaxation coefficient for rotation will be zero (infinite relaxation time).

Finally, the work of Fetter [6] on rectilinear vortices has an interesting bearing on the result obtained in (3.20), relating the local (macroscopic) rotation to the molecular angular velocity. He considered the circular cross section of the vortex core to be an assembly of "point" vortices (these are equivalent to the rotational nanoparticles referred to previously) in a lattice array, with rotational axes lying along the lattice axis direction. The array is specified by a set of two-dimensional position vectors $\{r_i\}$, and the total energy of the system is given by

$$E = \Sigma_i E_i + \left(\frac{1}{2}\right)\Sigma_{ij} E_{ij},$$
(3.31)

where E_i is the energy of the ith point vortex and E_{ij} is the interaction energy of the ith and jth vortex.

The interaction term is shown to lead to a uniform rotation of the array about its center. Transforming to a reference system rotating with angular velocity Ω, the total energy is then given by

$$H = E - \Omega L,$$
(3.32)

where L is the total angular momentum about the axis of rotation.

Fetter shows that the right-hand side of this equation is self-consistent only if the array (or core) angular velocity is related to the point vortex strength as follows:

$$\Omega = \left(\frac{1}{2}\right)nK,$$
(3.33)

where the discrete array has been replaced by a smoothed vortex density n, and K is the point vortex strength.

If the point vortices are considered as molecular particles of effective radius σ and angular velocity ω, then

$$K = 2\pi\sigma^2\omega,$$
(3.34)

and

$$\Omega = n\pi\sigma^2\omega. \tag{3.35}$$

This is completely analogous to the Barnett equation (3.20). The proportionality constant in each case is dimensionless, and, in the above equation, represents the total number of "molecular" vortices in the core cross section.

3.4 Transport and Optical Properties of the Core Gas

3.4.1 Heat Conduction

It has already been seen that the vortex core gas has zero shear and bulk viscosity.

Due to spin (or internal rotation) alignment, temperature differences can be leveled out by a wave propagation process rather than by a diffusion mechanism. It is, in effect, an internal convection mechanism and has been used to explain the anomalously high heat conductivity of superfluid liquid helium [6]. The spin-aligned gas can be regarded as a superfluid with respect to the normal state of the gas. Conversion to the normal state occurs by addition of heat to the spin-aligned state.

Based on this two-fluid concept, a wave equation for the gas temperature can be derived:

$$\left(\frac{c_p}{T}\right)\frac{\partial^2 T}{\partial t^2} - \left(\frac{\rho_s}{\rho_n}\right)s^2 \text{ div grad } T, \tag{3.36}$$

where c_p is the specific heat at constant pressure, ρ_s is the superfluid density, ρ_n is the normal fluid density, and s is the entropy density. A similar equation holds for the entropy density s.

The velocity of propagation of these thermal waves is given by

$$u^2 = (\rho_s/\rho_n)s^2 T/c_p. \tag{3.37}$$

Typical numerical values for the thermal wave velocity range around 5×10^3 cm/s.

For a steady state, London [7] derived the following expression connecting the temperature gradient and the heat flux **q**:

$$\Lambda \text{ grad } T = -\text{curl curl } \mathbf{q}, \tag{3.38}$$

where

$$\Lambda = (\rho_s)^2 T/\eta_n, \tag{3.39}$$

and η_n is the coefficient of shear viscosity of the normal fluid.

Equation (3.38) replaces the usual equation of heat conduction

$$\kappa \, \text{grad} \, T = -\mathbf{q}.$$

For a capillary of radius R, it can be shown that the effective coefficient of heat conduction is

$$\kappa \cong \frac{\Lambda R^2}{8}. \tag{3.40}$$

For a typical gas at atmospheric pressure and a value of $R = 0.1$ mm, this yields a value for the coefficient of heat conduction of

$$\kappa = 3 \times 10^{-2} \text{ cal/cm s°C}.$$

This is about a 100 times larger than the normal value for κ.

The high value for heat conduction in a vortex core is supported by the observation made by the author [22] that the flame speed in a vortex ring formed of premixed combustible gases is about five times higher than that to be expected in an irrotational gas at the same initial temperature and pressure. A 25-fold increase in the effective coefficient of heat conduction would explain this.

3.4.2 Optical Properties

Alignment of molecular angular momenta results in a net electric polarization in the vortex core fluid. It is the electric dipole moment induced by the incident electric dipole moment, or the electric permeability of the gas, which plays a central role in flow birefringence and the spectrum of depolarized Rayleigh scattering [8].

Let the electric dipole of a molecule be

$$\mathbf{u} = u_0 \mathbf{e}, \tag{3.41}$$

where

$$\mathbf{e}(t) = \mathbf{e}(0)\sin\omega_r t + \mathbf{l} \times \mathbf{e}(0)\cos\omega_r t,$$

ω_r is the rotational frequency, and $\mathbf{l}$ is the direction of the rotational angular momentum vector.

The time average of this dipole moment is zero. But the field $\mathbf{E}$ of the incident electromagnetic radiation changes this to a nonzero average. This *induced* dipole moment (for a linear nonpolar molecule) can be written as

$$\mathbf{u} = \langle\langle \alpha \rangle\rangle \cdot \mathbf{E}, \tag{3.42}$$

where the polarizability tensor $\langle\langle\alpha\rangle\rangle$ is related to $\mathbf{e}$ as follows:

$$\langle\langle\alpha\rangle\rangle = \alpha_{\text{para}}\mathbf{ee} + \alpha_{\text{perp}}(\langle\langle\delta\rangle\rangle - \mathbf{ee}). \tag{3.43}$$

α_{para} is the polarizability component parallel to the axis of symmetry and α_{perp} is the polarizability component perpendicular to the axis of symmetry.

The electric displacement vector is written as

$$\mathbf{D} = \langle\langle\varepsilon\rangle\rangle \cdot \mathbf{E}, \tag{3.44}$$

$$\langle\langle\varepsilon\rangle\rangle = \varepsilon\langle\langle\delta\rangle\rangle + \langle\langle e\rangle\rangle, \tag{3.45}$$

and

$$\langle\langle e\rangle\rangle = -2\pi n(\alpha_{\text{para}} - \alpha_{\text{perp}})\langle\langle\mathbf{ll}\rangle\rangle, \tag{3.46}$$

is the anisotropic (symmetric, traceless) part of the electric permeability tensor (3.48)

$$\varepsilon = 1 + \left(\frac{4}{3}\right)\pi(2\alpha_{\text{perp}} + \alpha_{\text{para}}), \tag{3.47}$$

is the dielectric constant.

Classically,

$$\langle\langle\mathbf{ll}\rangle\rangle = a_{\mu\nu}^{(2)}\left[(\sqrt{2})w_0^2/3\right], \tag{3.48}$$

where $a_{\mu\nu}^{(2)}$ is the polarization tensor.

The transport equation for $\langle\langle a\rangle\rangle$ in the field-free, shear free, vortex core simplifies to

$$\frac{\partial\langle\langle a\rangle\rangle}{\partial t} + \omega_T\langle\langle a\rangle\rangle = 0, \tag{3.49}$$

where $\omega_T \equiv \omega_{+2}^{(22)}$ is the relaxation coefficient for the tensor polarization.

It is this coefficient which determines the broadening of the depolarized Rayleigh line.

Hess [9, 10] has shown that the spectrum of the depolarized Rayleigh line is given by the Lorentzian:

$$S_{\text{L}}(\omega|\mathbf{k}) = \frac{\omega_T + k^2 D_T}{\{\omega^2 + (k^2 D_T + \omega_T)^2\}\pi}, \tag{3.50}$$

where $\mathbf{k}$ is the wave vector and

$$D_T = \frac{k_{\text{B}}T/m}{\omega_T}. \tag{3.51}$$

The relation between change in refractive index and the velocity gradient ∇u has been derived by Hess to be

$$\Delta n = \left[\frac{2\pi}{\sqrt{15}}\right](\alpha_{\text{para}} - \alpha_{\text{perp}})\left(\frac{\omega_{\eta T}}{\omega_T}\right)\left(\frac{\eta}{k_B T}\right)\nabla u, \qquad (3.52)$$

where $\omega_{\eta T} \equiv \omega_2^{(12)}$ is the relaxation coefficient for flow birefringence.

But in the vortex core, $\eta \equiv 0$ and so the gas in the vortex core is not birefringent. The gas in the vortex core has unique optical properties.

An experimental study of the spectrum of the depolarized Rayleigh scattered light is necessary to substantiate these predictions. Moreover, it should be possible to determine values of the relaxation coefficient ω_T and thus to gain information on the intermolecular force potentials.

3.5   Electric (Barnett) Polarization in the Boundary Layer

The author in [11] reports a kinetic theory for a dilute gas of rotating molecules in a vortex core. This predicted a polarization of the molecular rotations about the local rotation axis – a Barnett effect.

In the shear, or boundary, layer the streamlines are helical (not circular) and although a Barnett effect would be expected, there would be a nonzero shear and bulk viscosity and the gas *would* be birefringent. In at least two recent publications [12, 13], variable index of refraction fields has been observed in turbulent shear layers and traced to the presence of large vortical structures (with nonzero helicity) in the flow field.

Moreover, the author has made a very interesting observation on the interaction of vortex rings formed in air, with an external electric field. The vortices would not pass between the electrodes forming the field, but were deflected right or left, depending on the field polarity. This is understandable as the vortex core is a dipole which will interact with an electric field.

Vortex Electric Polarization: the vector polarization induced in the vortex core [9, 10] is

$$\mathbf{p} = \frac{(\sqrt{3})\langle\omega\rangle}{\omega_0} = \text{vortex induced dipole moment per unit volume}, \qquad (3.53)$$

where $\langle\omega\rangle$ is the mean angular velocity of the fluid particles in the core and

$$\langle\omega\rangle = \omega_0^2 \beta_0 I \omega, \qquad (3.54)$$

where

$$\omega_0 = \sqrt{(3k_B T/I)}.$$

k_B is Boltzmann's constant, T is the fluid temperature, and I is the moment of inertia of the vortex about the rotation axis:

$$\beta_0 = \frac{1}{k_B T}.$$

ω is the local (solid-body) angular velocity

3.6   Electric Charge–Dipole Interaction

In a boundary layer, there are many adjacent helical vortex cores, each of which is a dipole. These dipoles would interact with an external electric potential V, associated with some charge Q. Depending on the direction of rotation in the vortex cores and the polarity of the charge Q, this interacting force will be repulsive (pushing the boundary layer off the surface and decreasing drag) or attractive (pulling the boundary layer down on the surface and increasing drag).

For calculation purposes, Fig. 3.2 is a sketch of the interactive unit (charge–dipole). The force acting between the charge and dipole is

$$F = Q(1/4\pi\varepsilon_0)2p/d^3, \qquad (3.55)$$

where ε_0 is the permittivity of free space $= 9 \times 10^{-12}\ \mathrm{C^2/N\ m^2}$ and p is the dipole moment.

As the vortices in the boundary layer are co-rotating, all the dipoles have the same orientation.

The shear, or boundary, layer on a delta wing at Reynolds numbers such that the flow remains laminar will be considered here, as experimental data are available [14].

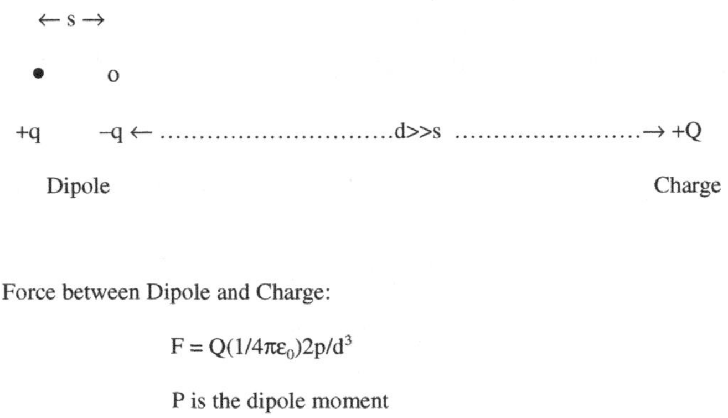

$\leftarrow s \rightarrow$

•   o

+q   –q $\leftarrow$ ....d>>s ....$\rightarrow$ +Q

Dipole   Charge

Force between Dipole and Charge:

$$F = Q(1/4\pi\varepsilon_0)2p/d^3$$

P is the dipole moment

Fig. 3.2   Electric charge–vortex dipole system

The force acting on all the dipoles in a volume V is

$$F_{tot} = Q(1/4\pi\varepsilon_0)2\mathbf{p}V/d^3, \tag{3.56}$$

where $\mathbf{p}$ is the vortex-induced dipole moment per unit volume – see (3.53).

Now $V = AL$, where A is the area of a region of the boundary layer and L is the vortex width. Substituting for $\mathbf{p}$ in (3.55) gives

$$F_{tot} = 6Q\left[\frac{\sqrt{I}}{k_B T}\right]\left(\frac{1}{4\pi\varepsilon_0}\right)\left(\frac{1}{d^3}\right)AL. \tag{3.57}$$

Hence, the "electric", or charge/dipole interaction, pressure on the boundary is

$$P_{QD} = F_{tot}/A = 6Q[(\sqrt{I}/k_B T)](1/4\pi\varepsilon_0)L/d^3. \tag{3.58}$$

It is convenient to convert the electric charge in coulombs, Q, into the electric potential, V in volts, and

$$V = \frac{\sigma}{2\pi\varepsilon_0}, \tag{3.59}$$

where σ is the charge per unit area $= Q/A$.

For $V = 100$ V, for example, $Q = 7.1 \times 10^{-8}$ coulombs.

3.6.1   Sample Calculation for the Boundary (Shear) Layer on a Delta Wing

I is the moment of inertia of the vortex about the rotation axis $= m/r^2$, where m is the mass of fluid in the cylindrical vortex and r is the radius of the cylinder.

Taking the vortex width as $L = 0.6$ cm and the length of the vortex as 2 cm, then I was calculated as $I = 4.95 \times 10^{-12}$ kg m^2.

A vorticity value of $\Omega = 700$ cm/s (Nelson and Visser) was used and a surface area of 4 m$^2 = A$.

The proposed boundary-layer control device is an insulated metallic layer about 10 cm below the boundary layer (upper edge of the airfoil) at a voltage V (volts) relative to "ground." A value of 100 V was used for this exploratory calculation.

The numerical calculation for the pressure on the boundary layer yielded

$$P_{QD} = 1.16 \times 10^{11}Q \text{ N/m}^2.$$

For $V = 100$ V, $Q = 7.1 \times 10^{-8}$

$$P_{QD} = 8.26 \times 10^3 \text{ N/m}^2.$$

The maximum pressure for separation would be the dynamic pressure of the flowing fluid,

$P_{sep} = (1/2)\rho U^2$ where U is the free stream velocity. For $U = 300$ m/s, for example, $P_{sep} = 2.7 \times 10^4$ N/m^2.

The P_{QD} pressure above is not quite enough to neutralize this. But, a voltage of 350 V would do so, and prevent separation under the "worst-case" scenario.

3.7 Concluding Remarks

Based on the kinetic theory for a dilute gas of rotating molecules developed by Hess and Waldman, it has been shown that polarization of the molecular axes occurs. Moreover, in the presence of local rotation (solid-body rotation), it has also been shown that polarization of the molecular rotational axes occurs along the local rotational axis – Barnett polarization.

The change in physical properties of the vortex core gas, which occurs as a result of Barnett polarization, is significant, and those related to the transport and optical properties of the gas have been dealt with. Such changes in the fluid physical properties are typical of those found in materials (solids and fluids) at the nanoscale level. There is a phase transition from the macroscopic fluid to the nanofluid.

Kinetic theory also predicts significant electric polarization $\mathbf{P}$ in the dielectric fluid flowing over an airfoil, where $\mathbf{P} = N\mathbf{p}$ and $\mathbf{p}$ is the average dipole moment per molecule. N is the number of molecules per unit volume.

The shear, or boundary, layer over a delta wing has been used as an example, where the flow is in the form of longitudinal helical vortices. Each helical vortex is a dipole and so the total dipole over any given area of the boundary layer can be computed.

Published physical values for a typical co-rotating boundary layer (vortex strength and vortex width) have been used.

The interactive force between this dipole assembly and an electrode charged to a voltage V buried in the airfoil is computed to the first order. As the force acts over an area, a pressure is determined. This can be an attractive pressure – pulling the boundary down on the airfoil surface, increasing the drag, but delaying separation. Alternatively, it can be repulsive, pushing the boundary layer off the airfoil surface, decreasing the drag and also the lift, and causing the airfoil to "slip" through the fluid. This condition has already been identified in the emerging area of fluid dynamics known as nanofluidics.

The implications for hypersonic and ionized fluid dynamics, with electric double-layer formation and the ensuing electrokinetics, are that alteration of hypersonic re-entry characteristics becomes feasible.

References

1. Barnett, S.J.: Phys. Rev. **66**, 224 (1944)
2. McCormack, P.D.: Proc. Roy. Irish Acad. **71A**(6), 73 (1971)
3. Waldman, L.: Zeits. Naturforsch. **15a**, 19 (1960)
4. Waldman L., Kupatt H.: Zeits. Naturforsch. **18a**, 86 (1963)
5. Steele, W.A.: J. Chem. Phys. **38**(10), 2404 (1963)
6. Fetter, A.L.: Phys. Rev. **162**, 143 (1967)
7. London, F.: Superfluids, vol. 2. Wiley, New York (1954)
8. Landau, L.D., Lifshits, E.M.: Electrodynamics and Continuous Media. Pergamon, Oxford (1960)
9. Hess, S.: Phys. Lett. **29A**, 108 (1969)
10. Hess, S.: Phys. Lett. **30A**, 239 (1969)
11. McCormack, P.D.: Proc. Roy. Irish Acad. **75A**, 57 (1975)
12. Oljaca, M.: Paper FC.06. Annual Meeting of the American Physical Society. New Orleans, LA (1999)
13. Dimotakis, P.: J. Fluid Mech. **433**, 105 (1999)
14. Nelson, R.C., Visser, K.D.: J. Am. Inst. Aero. Astro. **31**(1), (1993)

Chapter 4
Nanoduct Fluid Flow

4.1 Introduction

Nanoducts are unique in that fluid moving through them has intense vorticity, or molecular spin. In such regions of fluid flow, the molecular theory of fluids must be used to model the flow region and the physical properties of the fluid. Such properties change significantly in the presence of intense vorticity. These topics have been dealt with in Chap. 3 under the heading of the Nano-boundary Layer, using Waldman's kinetic theory for a fluid of rotating and translating molecules.

Expressions for the particle density, spin, and fluid velocity were derived (two particle collisions only). Allowing for macroscopic rotation of the gas led to an equation analogous to the Barnett equation. Expressions for the transport and optical properties of the nanoduct fluid were also derived.

4.2 Kinetic Theory for Fluid Transport Parameters

Given that the sound speed in the gas is given by

$$u_s = \sqrt{kRT},$$

(4.1)

where $k = c_p/c_v$ – the ratio of specific heats – R is the gas constant; only the important dimensionless numbers will be dealt with here.

$$\text{Mach number} = Ma = \frac{u}{u_s}.$$

(4.2)

A flow with Ma less than 0.3 can be treated as incompressible.

P. McCormack, *Vortex, Molecular Spin and Nanovorticity: An Introduction*,
SpringerBriefs in Physics, DOI 10.1007/978-1-4614-0257-2_4,
© Percival McCormack 2012

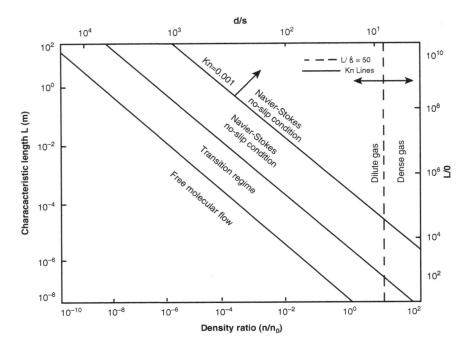

Fig. 4.1 Flow regimes for gases and relationship between the Knudsen number, characteristic length L, density n, and molecular distance.

$$\text{Reynolds number} = \frac{uL}{v}, \tag{4.3}$$

where L is the characteristic length scale of the flow, such as the channel diameter, and v is the kinematic viscosity of a gas.

The ratio between the mean free length, λ, and the characteristic length scale is called the Knudsen number

$$Kn = \frac{\lambda}{L} = \frac{\left(\sqrt{k\pi}/2\right)\text{Ma}}{Re}. \tag{4.4}$$

It can also be determined from the Mach number and the Reynolds number.

The Knudsen number is used to determine a suitable model to describe fluid flow in nanofluids.

For a small Kn value ($Kn < 10^{-3}$), the fluid is treated as a continuum with nonslip boundary conditions. For Kn values between 10^{-3} and 10^{-1}, the continuum model with slip boundary conditions are applied. For Kn values between 10^{-1} and 10, the flow is in the transition phase and can still be described using equations modified from the continuum model. For Kn values above 10, molecular dynamics can be used to describe the free molecular flow – see Fig. 4.1.

4.3 Molecular Dynamics and Monte Carlo Simulation

Molecular dynamics (MD) is a simulation method for the calculation of the motion of many particles in a system.

The interaction between the molecules in a system is described by Newton's second law and the simplest model of a molecule is a hard sphere with mass m. For two neutral molecules, the interaction can be described by the Lennard–Jones potential [1]:

$$\Psi_{ij}(r) = 4\varepsilon \left[c_{ij} \left(\frac{r}{\sigma} \right)^{-12} - d_{ij} \left(\frac{r}{\sigma} \right)^{-6} \right], \tag{4.5}$$

where r is the intermolecular distance, c_{ij}, d_{ij} are interaction coefficients, and σ is the molecule diameter and is the characteristic length.

The interaction force between two molecules is

$$F_{ij}(r) = -\frac{d\psi_{ij}(r)}{dr}. \tag{4.6}$$

The equation of motion is

$$\frac{d^2 \mathbf{r}_j}{dt^2} = \left(\frac{1}{m} \right) \sum_{j=1, j\neq 1}^{N} F_{ij}. \tag{4.7}$$

The basic steps of molecular dynamics simulation are listed as follows:

1. Determining the initial conditions and geometrical parameters.
2. Determining the interaction forces.
3. Integrating the equation of motion (4.7) for the next position (state) of the molecules.
4. Repeating over the required number of time steps.

The fluid model used in MD with Lennard–Jones interactions is called a "simple fluid" or Lennard–Jones fluid. This model forms a cost-effective model for describing liquid flow in the nanochannel. The continuum model with Navier–Stokes equations breaks down in a nanochannel below four molecular diameters, or approximately 1 nm.

Molecular dynamics is a deterministic method and so computational expense is very high. It is used with liquids. Problems with gases are solved by combining statistical methods and the deterministic method of particle dynamics. This approach is called DSMC, and instead of tracking the position of each single molecule, many molecules form a particle. The simulation of the particles is deterministic, while the interaction of the molecules in the particle is modeled statistically. There are three substeps in each step of DSMC: indexing and cross-referencing of the particles; modeling of particle motion; and simulation of collisions and probing macroscopic properties.

As in MD, the new position of the particle is determined by integrating the equation of motion over a time step Δt. To maintain the same number of particles in the simulation domain, particles exiting at one boundary should reenter at another boundary.

The simulation domain is divided into cells. Only particles in the same cell can collide and the cell size should be less than three times the mean free path. The probability of collision is then calculated for every particle pair. The method selects two arbitrary particles and calculates their relative velocity and if their relative velocity is above a certain threshold the pair is selected as a collision pair. The collision is then simulated and new velocities determined. The velocities of the particles after the collision are calculated based on the conservation of impulses and of kinetic energy. For the macroscopic properties, the average value over all particles is calculated. The temperature is evaluated from the kinetic energy of the particles. For evaluation of viscosity, see [2].

4.4   Diffusion in Nanochannels

In a nanochannel, the simplest mode of diffusive transport is Knudsen diffusion, where the channel size is smaller than the mean free path ($Kn > 1$). The diffusion is dominated by collision with the channel wall and not by collision with surrounding molecules or collision with surrounding molecules as estimated by the Stokes–Einstein theory.

For a liquid, the intermolecular distance is of the order of the molecular diameter, and so Knudsen diffusion is not significant. Under standard conditions, the mean free path of gas molecules is about 100 nm. Therefore, the diffusive transport of a gas through a nanochannel is mainly determined by Knudsen diffusion. The diffusion coefficient based on kinetic theory is

$$D_{Kn} = \frac{Lu_{\mathrm{p}}}{3}, \qquad (4.8)$$

where L is the characteristic length, such as channel diameter, and $u_{\mathrm{p}} = \sqrt{2RT}$ is the most probable molecular velocity [2, p. 48]. The Knudsen diffusion coefficient is therefore given by

$$D_{Kn} = \frac{L\sqrt{RT}}{3}, \qquad (4.9)$$

where $R = k/M$ is the gas constant, with M being the molecular weight.

4.5 Electrokinetics in Nanochannels

Surface charge on a channel wall is a result of dissociation of nonelectric adsorption of ions in the solution to the surface [3]. The surface charge can be positive or negative, depending on the pH of the solution. At a particular pH value, the net charge can be zero. For example, the surface charge of glass is zero at a pH value of about two [4]. The amount of charge per unit area is called the surface charge density:

$$\sigma_s = \frac{\sum_i z_i e}{A},$$ (4.10)

where z_i is the valence of ion, i.e., the elementary charge, and A is the surface area.

In nanochannels, counterions accumulate on the surface charge and form the Stern layer.

Outside this immobile layer, there is a thicker mobile layer – the Gouy–Chapman layer. Under an electric field, the latter layer can move relative to the solid surface. The immobile Stern layer and the mobile Gouy–Chapman layer form the electric double layer (EDL). The interface between these two layers is called the shear layer. These counterions extend into the channel through the Gouy–Chapman layer. Across the EDL, co-ions are repelled. The thickness of the EDL, the Debye length, is inversely proportional to the square of the bulk concentration – $\lambda_D \sim 1/\sqrt{n_i^0}$ – see [3, p. 24]. Thus, if the EDL is small compared to the channel, both ion types are transported through the channel, and the conductance is proportional to the ion concentration.

In a nanochannel, the surface charge density should balance the charge density in the solution:

$$\sigma_s = -\int_0^\infty \rho_e dx,$$ (4.11)

where the charge density is

$$\rho_e = \frac{-\varepsilon\varepsilon_0 d^2\psi}{dx^2}.$$ (4.12)

Using the Poisson–Boltzmann equation, the relation between the surface charge density σ_s and the surface potential ψ_s can be derived as [5]

$$\sigma_s = \sqrt{\left\{ 2\varepsilon\varepsilon_0 kT \sum_i n_i^0 \left[\exp\left(\left(\frac{-z_i e\psi_s}{kT}\right) - 1 \right) \right] \right\}}.$$ (4.13)

The simplified model for the conductance of a nanochannel for an electrolyte such as KCl($\mu_{K^+} = \mu_{Cl^-} = \mu_I, z = 1$) for all concentration is [6]

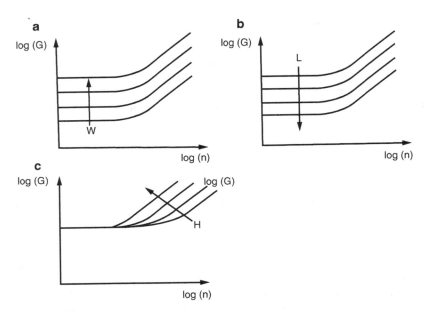

Fig. 4.2 Typical behavior of the conductance of a nanochannel; (**a**) dependency on channel width N, (**b**) dependency on channel length, L (**c**) dependency on channel height H, and dependency on conductance

$$G_{eo} = \left(\frac{2WH}{L}\right)\mu_i n_{KCl}e + \left(\frac{2W}{L}\right)\mu_i \sigma_s, \qquad (4.14)$$

where W, H, and L are the width, height, and length of the nanochannel, respectively.

Figure 4.2 shows the relation between the conductance of a nanochannel and the different parameters.

The change in width and length follows the rule of an ohmic conductor [the conductance is proportional to the width and inversely proportional to the length (Fig. 4.2a, b)].

4.6   Slip Flow in Nanoducts

Initially compressible (gas) flows will be dealt with. In the Knudsen number range $-0.01 \leq Kn \leq = 0.1$, the often-assumed no-slip boundary condition appears to fail and a sublayer of the order of one mean free path, known as the Knudsen, starts to become dominant between the bulk of the fluid and the wall surface. The flow in the Knudsen layer cannot be analyzed with the Navier–Stokes equations and it requires special solutions of the Boltzmann equation. However, for $Kn \leq 0.1$, the Knudsen layer covers less than 10% of the channel height and this layer can be

Fig. 4.3 Slip length, L_s, for a simple shear flow along a flat plate

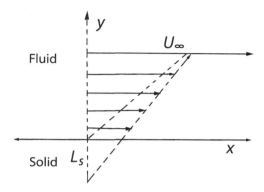

neglected by extrapolating the bulk gas flow toward the wall. This results in a finite velocity slip value at the wall, and this flow regime is known as the *slip flow regime*. In this slip flow regime, the flow is governed by the Navier–Stokes equations modified by Maxwell's velocity slip at the boundary.

In 1879, Maxwell used the kinetic theory of gases to identify the slip boundary condition,

$$u_x(x, y) = \left(\frac{2 - \text{TMAC}}{\text{TMAC}} \right) \wp \left[\frac{\partial u_x(x, y)}{\partial y|_w} \right], \tag{4.15}$$

where $\wp$ is the mean free path $\sim 1/d^2 n$, d is the effective molecular diameter, and n is the molecule number density.

TMAC is the tangential momentum accommodation coefficient, frequently of value one. Therefore, the fluid velocity at the solid surface is assumed to be proportional to the shear rate at the surface, and the proportionality constant has dimensions of length.

The slip length, L_s, is defined in terms of the accommodation coefficient and the mean free path as

$$L_s = \left(\frac{2 - \text{TMAC}}{\text{TMAC}} \right) \wp, \tag{4.16}$$

and also can be defined [7] as the ratio of the slip length to the surface shear rate:

$$u_x(x, 0) = L_s \left[\frac{\partial u_x(x, 0)}{\partial y} \right]. \tag{4.17}$$

In physical terms, the slip length is the distance inside the solid where the fluid velocity extrapolates linearly to zero (Fig. 4.3).

Several theories for the generation of slip have been developed and these include (1) surface roughness [4] and (2) nano-bubbles on the surface and applied shear [5]. Shear is probably the most interesting. It has been shown [6] that at low shear rates a linear boundary condition exists. After a critical value of the shear rate γ'_C, Thompson and Troian [6] suggest that the relation between the slip length and the shear rate is nonlinear where

$$L_s = L_s^0 \left(1 - \gamma'_C \left[\frac{\partial u_x(x,0)}{\partial y} \right] \right)^{-1/2}, \tag{4.18}$$

with L_s^0 and γ'_C being constants and being particular to the materials involved. In [5] it is shown that for low and moderate shear rates, the slip length is not affected.

In [7] a nonlinear slip boundary condition

$$u_x(x,0) = L_s \left\{ \left(\frac{\partial u_x(x,0)}{\partial y} \right)^n \right\}, \tag{4.19}$$

is used to solve for velocity and for various slip lengths, using a reduced form of the boundary layer equations. Results for $n = 1/2$, corresponding to a convergent channel, and $n > 1/2$, corresponding to flow past a wedge, are determined. The case of $n = 1$ is invalid from the arguments in [7] since the solution for one of the constants of integration would not exist for $n = 1$.

4.7   Water Flow in Nanochannels

Water has many unique properties and has many nanofluidic applications and so its behavior in nanochannels will be dealt with.

A water molecule has two hydrogen atoms and one oxygen atom – see Fig. 4.4. The hydrogen atoms are linked to the oxygen via two bonds with a length of 0.970 Å

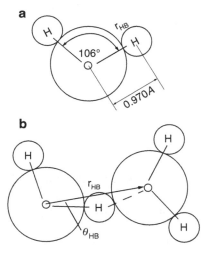

Fig. 4.4

each. The two bonds form an angle of 106°. Thus, the water molecule acts as an electric dipole, where the oxygen site appears negatively charged and the hydrogen site positively. The symmetry axis of the water molecule is the dipole axis. The dipole moment of a water molecule in the liquid state at 300 K has been determined experimentally as 1.95 ± 0.2 Debye (1 Debye $= 3.336 \times 10^{-30}$ cm) – [8].

The main problem for MD simulation involving water is the accurate description of the interaction potential between molecules. For example, there are 40 different models for the simulation of water. In addition to the Lennard–Jones potential, electrostatic interactions of the dipole are considered for simulating the interaction between the molecules. Each model is optimized to fit a critical parameter, but cannot be applied in general.

A key issue of research on water in the nanoscale is the nature of the hydrogen bond (HB) network. The strength of about 21 kJ/mol of a hydrogen bond is between that of a covalent chemical bond (~420 kJ/mol) and that of the weak van der Waals interaction (~1.25 kJ/mol). Energy and geometric conditions are used to determine the formation of an HB. The energy condition is based on the interaction energy. If the interaction energy between two molecules is below a threshold energy (about −10 kJ/mol), an HB can be formed. The geometric condition is based on the distance and angle between the bonds of the atoms (see Fig. 4.4). If the distance between two oxygen atoms (the first coordination shell) and the angle between the O–O and O–H bonds (30°) are less than a certain threshold, an HB can be formed.

As HBs determine the properties of bulk water, the influence of nanochannels on HB determines the type of the transport. A highly oriented HB has been reported (see [9]) inside a carbon nanotube. An HB is much more stable inside the channel. The average lifetime of an HB inside a carbon nanotube is 5.6 ps, while it is 1.0 ps in bulk water [9]. At a critical channel diameter of 8.6 Å, water molecules are immobilized in the nanotube as a stable HB network [10]. This so-called water wire allows protons to move from one molecule to another resulting in the selective proton-conducting properties of the nanotube. The transport of the "water wire" in a nanotube is as follows (see [11]). A water molecule enters the channel with the hydrogen atom first. Water fills the channel in a chain form with the same orientation of water molecules. Depending on the interaction with the channel, the chain can be broken, resulting in a burst of molecule transport.

Molecular dynamics (MD) is suitable for studying transport phenomena in subnanometer channels such as the carbon nanotube. It has been found [12] with MD simulation that inside a CNT with a diameter of 0.8 nm, the number of hydrogen bonds of water decreases with respect to bulk water in larger channels. Water molecules align in a one-dimensional line and can move quicker through the CNT. In contrast to larger channel flows, water transport through subnanometer channels is semi-frictionless and nearly independent of the channel length. MD simulation can be combined with the continuum approach to avoid the very large computational expense. For example, the near-wall region can be simulated with MD, while the bulk area is solved with the continuum model. The method is suitable for a relatively large nanochannel of about 10–100 nm. This concept has been applied [13] to Couette and Poiseuille flow where the near-wall region should

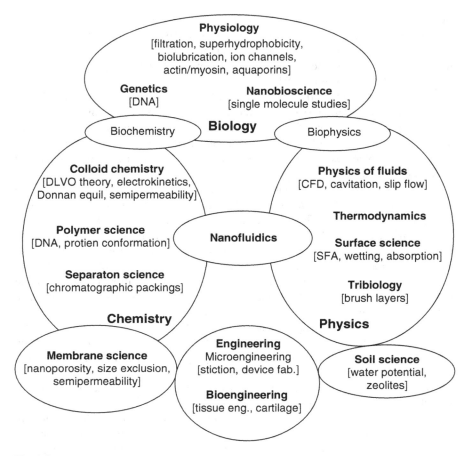

Fig. 4.5

be at least 12 molecular diameters in size. The continuum and MD domains should overlap by at least ten molecular diameters.

It is appropriate at this stage to summarize the disciplines related to nanofluidics and related scientific subjects (see Fig. 4.5).

4.8 Molecular Spin in Nanoduct Fluid Flow

In nanofluidic devices, very small volumes of fluid are transported around and mixed in nanosized channels and chambers, and the flow frequently driven by pumping and mixing mechanisms which generate nonzero mean oscillatory flows [14].

Nanofluidic flows are characterized by many distinct features such as large surface to volume and low Reynolds number [15]. On these small scales, it is doubtful whether the Navier–Stokes description is valid.

It has been shown [16] using molecular dynamics simulations of a Poiseuille flow that the classical Navier–Stokes theory is satisfactory for systems with a characteristic length scale down to 7–8 atomic diameters. Data from molecular dynamics simulations have compared well with the solution to the Navier–Stokes equation of nano-confined fluids in oscillatory flow with frequencies up to 10^2 GHz [17].

But the Navier–Stokes theory ignores several microscopic degrees, assuming that they have no effect on the translational fluid motion. However, it is well known that for molecular fluids the translational momentum couples to the intrinsic angular momentum via an exchange between the fluid vorticity and the molecular angular velocity [18]. Hansen et al. [19] have derived the extended Navier–Stokes equations (including the effect of molecular spin) for isotropic and dense fluids.

4.8.1 Extended Navier–Stokes Equations

These include the coupling between the translational velocity field $\mathbf{u}$ and the spin angular momentum per unit mass $\mathbf{s}$:

$$\frac{D\rho}{Dt} = -\rho(\nabla \cdot \mathbf{u}), \tag{4.20}$$

$$\frac{\rho D\mathbf{u}}{Dt} = -\nabla p + \left(\eta_v + \frac{\eta_o}{3} - \eta_r\right)\nabla(\nabla \times \mathbf{u}) + (\eta_o + \eta_r)\nabla^2\mathbf{u} + 2\eta_r\,(\nabla \times \mathbf{\Omega}), \tag{4.21}$$

$$\frac{\rho D\mathbf{s}}{Dt} = 2\eta_r(\nabla xu - 2\mathbf{\Omega}) + \left(\zeta_v + \frac{\zeta_o}{3} - \zeta_r\right)\nabla(\nabla \cdot \mathbf{\Omega}) + (\zeta + \zeta_r)\nabla^2\mathbf{\Omega}, \tag{4.22}$$

where ρ is the mass density, p is the pressure, and $\mathbf{\Omega}$ is the spin angular velocity. The transport coefficients η_v, η_o, and η_r are the bulk, shear, and rotational viscosities, respectively, and ζ_v, ζ_o, and ζ_r the equivalent vortex spin viscosities. Note that the total angular momentum is the sum of the orbital angular momentum and the spin angular momentum.

For a dense-enough incompressible fluid, these equations can be approximated to

$$\frac{\rho D\mathbf{u}}{Dt} = -\nabla p + (\eta_0 + \eta_r)\nabla^2\mathbf{u} + 2\eta_r(\nabla \times \mathbf{\Omega}). \tag{4.23}$$

$$\frac{\rho D\mathbf{s}}{Dt} = 2\eta_r(\nabla \times \mathbf{u} - 2\mathbf{\Omega}). \tag{4.24}$$

In a fluid where the divergence of the velocity and angular velocity fields are zero, it can be shown [19] that

$$\nabla^2 \Omega = \left(\frac{1}{2}\right) \nabla^2 \omega. \tag{4.25}$$

where ω is the vorticity and is decoupled from the angular velocity.

In general, the contribution to the total dissipation from the translational motion is always positive and orders of magnitude greater than that from the spin. Hence, the total dissipation is only slightly affected by the molecular spin.

4.9 Nanoscale Forces

These are exerted by walls on particles or solvent molecules or on nearby walls, or by the particles and molecules on each other. These forces govern the behavior of the molecules or particles in nanostructures. They give rise to equilibrium phenomena, such as differences in ionic distribution, or to kinetic phenomena such as (macroscopic) viscosity. Figure 4.6 is a schematic of a number of forces acting in the nanoscale, for the case of interaction between a spherical particle representing an AFM tip and a flat surface [20]. It is seen that the spatial extension of the forces is maximum for the electrostatic forces and minimum for the van der Waals forces. Electrostatic forces act as far as the electrical double layer extends – typically from 1 to 100 nm, depending on the electrolyte concentration and are repulsive or attractive – while van der Waals predominate at distances less than 2 nm and are always attractive.

A very useful tool in the study of nanoscale forces is the surface forces apparatus (SFA) – see [21]. It also allows salvation forces to be observed and quantified – these forces provide attachment of layers of solvent molecules to the surface. Also, hydration forces can be observed, occurring when ionic surface sites or ions in the liquid resist dehydration [22]. In biology and medicine, joint lubrication is an important area of research and hydration forces are involved [23].

A strong force is the capillary force and SFA is a useful tool here also [24].

So far, only forces that are normal to the surface of the SFA, important in static conditions, have been considered. However, shear forces on the nanoscale have also been shown to be different from the macroscopic domain. For simple Newtonian liquids, friction is seen to increase from the macroscopic value when the separation between two surfaces becomes less than about ten molecular layers. A change in film properties from liquid to solid-like behavior is often seen, occurring, for example, for alkanes but not for alcohols [25]. The shape of the molecule appears to determine this. Symmetric-shaped molecules order themselves in neat layers and turn into a quasi-solid, but asymmetrical ones remain in a disordered liquid state [26]. Note that for the experiments mentioned, atomically smooth surfaces are used, in contrast to most experiments where surfaces will be rough.

Fig. 4.6

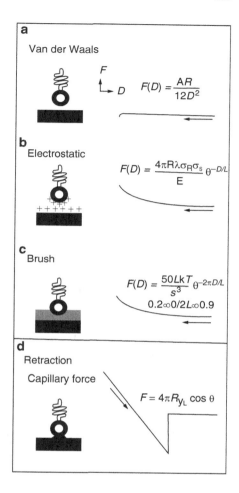

Surface roughness is one of the reasons for the hysteresis, often observed in approach/separation processes such as wetting/dewetting, adhesion/decohesion, and advance/receding of the contact angle [27]. Another important issue in nanofluidics is that, sometimes, the liquid is observed to slip past the surface decreasing the fluid resistance [28]. As dealt with earlier in this chapter, when this occurs the commonly used boundary condition is no longer valid and a so-called slip length has to be introduced.

Atomic force microscopy (AFM) has become the best method for measuring surface, because it enables measurements to be made quickly and with high spatial resolution [20]. It has, however, a lower sensitivity than SFA (see review by Froberg et al. [29]).

Forces of lesser importance on the nanoscale are the gravitational and inertial forces.

Fig. 4.7 Optical micrograph
of a water plug (*dark color*)

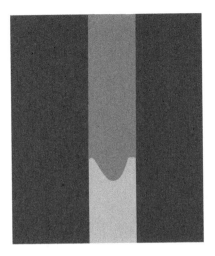

4.10   Theoretical Analyses

The forces between the individual atoms ultimately determine the behavior of any
system. The best model will be the one which accounts for all the individual atomic
interactions. Available computational power limits the system size to about 10 nm;
however, above this size systems can often be described using continuum theory,
which statistically averages the single interactions. For the behavior of a liquid
confined in a space below ten molecular diameters, where molecular layers start
being quantized, deviations from the predictions of classical continuum theory have
been observed. Molecular dynamics (MD) simulations are appropriate for such
systems [30]. Between 10- and 100-nm-sized channels, the lattice Boltzmann
equation has been very useful for the descriptions with complicated geometry and
composition [31].

4.11   Overview of Phenomena Occurring at the Nanoscale

4.11.1   Surface-Energy-Related Phenomena

At nanometer dimensions, the surface to volume ratio increases and surface-related
phenomena become increasingly dominant. For example, during two-phase flow in
nanochannels, the generation of capillarity-induced negative pressure is observed
[32]. This is impossible in gases, but liquids and solids can be put under tension –
they can exist at negative pressures [33]. Such pressures have been observed in
devices where the thin channel roofs were deformed by the negative pressure in a
capillary liquid plug. Figure 4.7 shows the tongue-like shape of the meniscus

observed in 100-nm-high nanochannels due to the bending of the 8.30-nm-thick and 10-μm-wide silicon nitride-capping layer under the influence of the negative liquid pressure. The pressure has been calculated at a value of -17 bar [34].

Earlier it has been mentioned that slip flow can occur at the (sub) nanometer scale. Slip can occur both at atomically flat surfaces, e.g., at high shear rates [35], and at hydrophobic rough surfaces where air pockets can remain [28] or a very thin vapor layer can adhere [36]. MD simulations of the movement of single water files through (6,6) carbon nanotubes indicate that slip also can occur at this scale because the water molecules interact more strongly with each other than with the walls [37]. The flow resistance in these simulations was independent of nanotube length between 1 and 4.5 nm.

4.11.2   Shear-Related Phenomena

Strong shearing, as mentioned earlier, can cause cavitation, either at the liquid/solid interface or in the liquid itself. Shear forces can also extend polymer and DNA molecules [38]. Long DNA can easily be broken and proteins denatured [39].

4.11.3   Electric Double-Layer-Related Phenomena

These occur on double-layer overlap on nanochannels or nanoslits. They are observed when studying nanochannels, nanopores, membranes, sol gels, and soils. Due to the large surface to volume ratio, the contribution of the surface conductivity (the conductivity through the electrical double layer) becomes larger at the nanoscale [40, 41].

When double-layer overlap occurs, the streaming potential will decrease and the pore conductivity will increase [42]. Moreover, co-ions will be excluded from the channel, or membrane; counterions are enriched and a Donnan equilibrium will be established between the channel solution and its connected reservoirs [42]. When double-layer overlap occurs in sols and gels, they swell due to the osmotic pressure resulting from the high concentration of counterions. This effect also contributes to the mechanical strength of cartilage [43]. Hydrostatic pressure will, therefore, be built up in nanochannels on double-layer overlap.

In the kidney, the negatively charged albumin is prevented from passing from the blood to the primary urine by the basal membrane. Co-ion exclusion by the negatively charged membrane contributes to this effect [44]. Similarly, the wall charge on nanotubes influences the transport of proteins with different charges [45]. Semipermeability has also been demonstrated in nanochannels using fluorescent molecules [46]. In semipermeable membranes with smaller pores, the effect is used for desalination [47].

4.11.4 Entropy-Related Phenomena

A system can be driven in a certain direction by an entropy increase. On the nanoscale, several phenomena have been shown to relate to entropy changes. Flexible molecules, such as DNA, have many more possible coiled states than elongated ones. Entropy, therefore, drives them to a coiled state [48]. Hence, DNA molecules will tend to stay in larger cavities instead of smaller ones, to maximize entropic gain [49].This will affect DNA separation in gels and in nanomachined entropic traps [50]. It will also drive DNA out of small cavities if it gets caught. DNA can be stretched by confinement in 2-D nanochannels if the diameter of the channel is less than the persistence length, which is about 50 nm [51].

4.11.5 Molecular-Structure-Related Phenomena

A major challenge for nanofluidics is to harness interactions on the single molecular scale. This is used in pharmacology where drugs are designed to fit in a certain receptor cavity. The drugs then have their effect on the human body by a massively parallel action in many cells.

Aquaporins conduct water molecules but not protons, which probably results from the intricate electrical and spatial properties of the channel, concerning interactions of single water molecules with the amino acids forming the walls [52]. Water conductance through a single aquaporin is about 10^9 molecules per second, and one could imagine a structure with many channels in parallel, conducting water but not protons. Manufacture of functional nanopores is still in its infancy.

4.12 The Field of Nanofluidics

This is often defined as the application and study of fluid flow around and inside nanoscale objects. It has its origins in many disciplines, and the phenomena encountered involve many areas in physics, chemistry, and biology.

Separation science is one area that has already been, and will be more so in the near future, greatly impacted. There are two basic reasons for this: first, the domain has been reached where the analyte size matches the device features and, second, absolutely regular features can now be made, which repetitively perform exactly the same operation on the particles to be separated.

A second area of nanofluidic applications is the study of the fundamental properties of liquids and molecules in, for example, fluid mechanics and biophysics. Will it be possible to apply massively parallel nanofluidic systems for computational purposes? Will it be possible to make structures that write DNA and read it?

Due to the nanotech tools developed recently, it is now possible to control liquid flow and molecular behavior at the nanoscale.

The new area of nanofluidics has already produced exciting results and more are expected.

References

1. Israelachvili, J.: Intermolecular and Surface Forces. Academic, New York (1992)
2. Bird, G.A.: Molecular Gas Dynamics and the Direct Simulation of Gas Flows. Clarendon, Oxford (1994)
3. Behrens, S.H., Grier, D.G.: The charge of glass and silica surfaces. J. Chem. Phys. **115**, 6716 (2001)
4. Priez, N., Troian, S.M.: Influence of wall roughness on slip behaviour. J. Fluid Mech. **554**, 25–46 (2006)
5. Huang, P., Breuer, K.S.: Direct measurement of slip length in electrolyte solutions. Phys. Fluids **19**, 028104 (2007)
6. Thompson, P., Troian, S.M.: A general boundary condition for liquid flow at solid surfaces. Nature **389**, 360–362 (1997)
7. Matthews, M.T., Hill, J.M.: Nano boundary layer equation with non-linear navier boundary condition. J. Math. Anal. Appl. **333**, 381 (2006)
8. Gubskaya, A.V., Kusalik, P.G.: The total molecular dipole moment for liquid water. J. Chem. Phys. **117**, 5290–5302 (2002)
9. Hummer, G., Rasaiah, J.C., Noworyta, J.P.: Water conduction through the hydrophobic channel of a carbon nanotube. Nature **414**, 188–190 (2001)
10. Mashl, R.J.: Anomalously immobilized water: a new phase induced by confinement in nanotubes. Nano Lett. **3**, 589–592 (2003)
11. Waghe, A.J.C., Rasaiah, J.C., Hummer, G.: Filling and emptying kinetics of carbon nanotubers in water. J. Chem. Phys. **117**, 10789–10795 (2002)
12. Gordillo, M.C., Marti, J.: Hydrogen bond structure of liquid water confined in nanotubes. Chem. Phys. Lett. **329**, 341–345 (2000)
13. Yen, T.H., Soong, C.Y., Tzeng, P.Y.: Hybrid molecular dynamics continuum simulation for nano/mesoscale channel flows. Nanofluidics **3**, 665–675 (2007)
14. Hansen, J.S., Ottesen, J.T.: Molecular simulation of oscillatory flows in microfluidic channels. Microfluid. Nanofluidics **2**, 301 (2006)
15. Bruus, H.: Theoretical Microfluidics. Oxford University Press, New York (2008)
16. Travis, K.P., Gubbins, K.E.: Poiseuille flow in narrow slit pores. J. Chem. Phys. **112**, 1984–1994 (2000)
17. Hansen, J.S., et al.: Local linear viscoelasticity of confined fluids. J. Chem. Phys. **126**, 144706 (2007)
18. de Groot, S.R., Mazur, P.: Non-equilibrium Thermodynamics. Dover, Mineola (1984)
19. Hansen, J.S., Daivis, P.J., Dodd, B.D.: Molecular spin in nano-confined fluidic flows. Microfluid. Nanofluidics **6**, 785–795 (2009)
20. Heinz, W.F., Hoh, J.H.: Spatially resolved force microscopy of biological surfaces using the AFM. Trends Biotechnol. **17**, 143–150 (1999)
21. Derjaguin, B.V., et al.: Investigations of the forces of interactions of surfaces in different media and the problem of colloid stability. Discuss. Faraday Soc. **18**, 24–41 (1954)
22. Raviv, U., Klein, J.: Fluidity of bound hydration layers. Science **297**, 1540–1543 (2002)
23. Raviv, U., et al.: Lubrication by charged polymers. Nature **425**, 1540–1543 (2003)
24. Wanless, E.J., Christenson, H.K.: Interaction between surfaces in ethanol. J. Chem. Phys. **101**, 4260–4267 (1994)
25. Mugele, F., Salmeron, M.: Frictional properties of thin chain alcohol films. J. Chem. Phys. **114**, 1831–1836 (2001)

26. Israelachvili, J., et al.: Liquid dynamics in molecularly thin film. J. Phys. Condens. Matter **2**, SA89–SA98 (1990)

27. Israelachvili, J.N., et al.: Dynamic properties of molecularly thin films. Science **240**, 189–191 (1988)

28. Vinogradova, O.I.: Slippage of water over hydrophobic surfaces. Int. J. Miner. Process. **56**, 31–60 (1999)

29. Froberg, J.C., et al.: Surface force and measuring techniques. Int. J. Miner. Process. **56**, 1–30 (1999)

30. Schoen, M., et al.: Shear forces in molecularly thin films. Science **245**, 1223–1225 (1989)

31. Succi, S.: The Lattice Boltzman Equation. Oxford University Press, Oxford (2001)

32. Tas, N.R., et al.: Capillarity induced negative pressure of water plugs in nanochannels. Nano Lett. **3**(11), 1537–1540 (2003)

33. Imre, A., Martinas, K., Rebelo, L.P.N.: Thermodynamics of negative pressures in liquids. J. Non-equilib.Thermodyn. **23**(4), 351–375 (1998)

34. Mercury, L., et al.: Thermodynamic properties of solutions in metastable systems. Geochim. Cosmochim. Acta. **67**(10), 1769–1785 (2003)

35. Zhu, Y., Granick, S.: Rate-dependent slip of newtonian liquid at smooth surfaces. Phys. Rev. Lett. **87**(9), 096105 (2001)

36. de Gennes, P.-G.: On fluid/wall slippage. Langmuir **18**, 3413–3414 (2002)

37. Kalra, A., Garda, S., Hummer, G.: Osmotic water transport through carbon nanotube membranes. Proc. Natl. Acad. Sci. U.S.A. **100**(18), 13770–13773 (2003)

38. Bakajin, O.B., et al.: Electrodynamic stretching of DNA in confined environments. Phys. Rev. Lett. **80**(12), 2737–2740 (1998)

39. Bao, G.: Mechanics of biomolecules. J. Mech. Phys. Solids **50**(11), 2237–2274 (2002)

40. Lykema, J.: Surface conduction. J. Phys. Condens. Matter **13**(21), 5027–5034 (2001)

41. Stein, D.: Surface-charged-governed ion transport in nanofluidic channels. Phys. Rev. Lett. **93** (3), 035901 (2004)

42. Fievet, P., et al.: Evaluation of 3 methods for the characterization of membrane-solution interface. J. Membr. Sci. **168**(1–2), 87–100 (2000)

43. Sun, D.D.: The influence of the fixed negative charges on mechanical and electrical behaviours in articular cartilage under unconfined compression. J. Biomech. Eng. **126**(1), 6–16 (2004)

44. Bethansen, L., et al.: Plasma disappearance of glycated and non-glycated albumin in diabetes mellitus. Diabetologia **36**(4), 361–363 (1993)

45. Ku, J.R., Stroeve, P.: Protein diffusion in charged nanotubes: on-off behaviour of molecular transport. Langmuir **20**(5), 2030–2032 (2004)

46. Pu, Q.S., et al.: Ion-enrichment and ion-depletion effect of nanochannel structures. Nano Lett. **4**(6), 1099–1103 (2004)

47. Probstein, R.F.: Physicochemical Hydro-dynamics: An Introduction, 2nd edn. Wiley, New York (1994)

48. Strick, T., et al.: Twisting and stretching single DNA molecules. Prog. Biophys. Mol. Biol. **74** (1–2), 115–140 (2000)

49. Muthukumar, M., Baumgartner, A.: Effects of entropic barriers on polymer dynamics. Macromolecules **22**, 1937–1946 (1989)

50. Han, J., et al.: Entropic trapping and escape of long DNA molecules. Phys. Rev. Lett. **83**(8), 1688–1691 (1999)

51. Tegenfeldt, J.O.: Stretching DNA in nanochannels. Biophys. J. **86**(1, pt 2), 596A (2004)

52. Tajkhorshid, E.: Control of the selectivity of the aquaporin water channel family by global orientational tuning. Science **296**(5567), 525–530 (2002)

53. Iler, R.K.: The Chemistry of Silica. Wiley, New York (1979)

54. Schoch, R.B., Han, J., Renaud, P.: Transport phenomena in nanofluidics. Rev. Mod. Phys. **80**, 839–883 (2008)